MW01593815

The MultiThread Marketer

How To Hire One (Or Better Yet) Become One

Douglas E. Mitchell

Published by FastPencil, Inc.

Published by FastPencil, Inc.
3131 Bascom Ave.
Suite 150
Campbell CA 95008 USA
(408) 540-7571
(408) 540-7572 (Fax)
info@fastpencil.com
http://www.fastpencil.com

Please note that much of this publication is based on personal experience and anecdotal evidence. Although the author has made every reasonable attempt to achieve complete accuracy of the content in this book, he assumes no responsibility for errors or omissions.

Also, you should use this information as you see fit, and at your own risk. Your particular situation may not be exactly suited to the examples illustrated here; in fact, it's likely that they won't be the same, and you should adjust your use of the information and recommendations accordingly.

Any trademarks, service marks, product names or named features are assumed to be the property of their respective owners, and are used only for reference. There is no implied endorsement if we use one of these terms.

Finally, Nothing in this book/ebook is intended to deliver legal advice. Always seek the services of professionals. This book/ebook was written to inform and entertain the reader.

Sharing this Document
There was a lot of work that went into putting this book/ebook together. I can't tell you how many countless hours went into this project. That means that this information has value, and your friends, neighbors, and co-workers may want to share it.

The information in this document is copyrighted. I would ask that you do not share this book/ebook with others-you purchased this book, and you have a right to use it. Another person who has not purchased this book does not have that right. It is the sales of this valuable information that makes the continued publishing other works possible.

If your friends think this information in this book/ebook is valuable enough to ask you for it, they should think it is valuable enough to purchase on their own. After all, the price is low enough that just about anyone should be able to afford it.

The Publisher makes no representations or warranties with respect to the accuracy or completeness of the contents of this book and specifically disclaim any implied warranties of merchantability or fitness for a particular purpose. Neither the publisher nor author shall be liable for any loss of profit or any commercial damages.

First Edition

This book is dedicated to everyone with the courage to ASK.

❧

Steve:

F T Rocks the

Marketing House!

Thanks for all the help

through the years...

701

Acknowledgements

It's always an awesome undertaking to extract the thoughts, nuggets, rants, learning, and neuroses from one's gray matter and put in ink (or 1's and 0's). Finding the time is always the key as the desire and passion burn hot every single day.

But to make a "book" happen no matter the form, it takes many others willing to put up with your almost maniacal idea testing, proofreading, etc. To all of you I say…deal with it because it will happen again. Kidding?

To my wife Steph without whom I'd wander aimlessly: I love you and thank you for putting up with my constant blathering about this or that. You are a modern day Superhero Chick and yes, you deserve many more girl weekends. A&4E

To GV and GA: Every day I learn more from you. I now understand how a look in your eyes and a simple hug is worth more than all the treasures in the universe. Fist bump.

To Mom and Dad: Sorry calls have been slim lately but as a President may have once said, "Writing a book is hard". You both gave me the ability to tell the world exactly how I was going to play with it.

To Michael C. Wagner: I thank you so much for offering the military grade flame thrower brain power that reignited my passion and fire for consuming knowledge at a pace that mere mortals find nauseating at best. I am smarter and better because of you and your Renaissance approach to leadership, business, and critical thinking. Stay deep. (www.michaelcwagner.com)

To Dan Pink, Clay Shirky, Nick Carr, Stephen Blank, Mihaly Csikszentmihalyi, Dan Ariely, and dozen more authors: You all have helped me understand my brain and why and how it operates. Ultimately this will help me help others understand how to make better decisions.

To Marc Sampson and John von Harz: You guys saw that Multithread Marketing was exactly what IPG needed to transform into BirdDog and that I would be a valuable part of the management team to execute. We both won.

To the State of Iowa: Your awesomeness knows no bounds. You occasionally abuse me with your frigidity or humidity (yikes) but you have given The Mitchell's the life of Riley whoever that is. Now that Trader Joe's is open and BirdDog moved 4 minutes from home, you've pretty much solidified your position as the "Hassle Free Capitol" of the Universe.

To The Brand Chef, Andrew Clark: You helped me fulfill my goals by buying my company and providing such excellent work through the years. You are most awesome and going out on a limb, believing in what I'd built, and taking it in new super cool directions has been a pleasure to watch. (www.createwowmarketing.com)

To Mike Sansone: You showed me in 2005 that blogging wasn't a toy but a business tool. It's quite possible that you reaching out to me by finding me writing about Iowa through some RSS feed alert was the catalyst to this moment in time. Thank you BlogFather. (www.conversta-tions.com)

To Andy Brudtkuhl: You have provided the hybrid tech-business mojo that gave me the ability to execute for my clients and myself. From our first ILE to tweaking out landing pages for ORS you really are "The Man". (www.48web.com)

To Adam Steen: From a 6AM coffee meeting to happy hours that last 6, thanks for being a sounding board, idea man, and connector of dots. Basement offer still stands. (www.25connections.com)

To Cathy Prince: Your editing skills have hopefully pre-vented me from looking like an uneducated rube. You went above and beyond the call of duty and performed with excellence. Now world, email her and give her your projects. (cjprince@q.com)

Contents

Preface

Hey, Doug Mitchell here. Let me start by saying I'm absolutely NOT an authority on hiring marketers. In fact, *I haven't hired a single marketer in my 16 professional years.* Does that disqualify me from writing this book? Read on and find out.

Doug Mitchell Livestreaming at a Conference for Entrepreneurs

I'm 39 and have been a Marketing Director and VP of Marketing over the last 11 years in start-up and early-stage companies. That means I made up my titles and did just about everything in those companies except hard-core coding and bookkeeping. (Isn't start-up life great?)

Start-up life requires agility and willingness to be uncomfortable. You'll learn way more about topics you never imagined existed, you'll fail, change course, and need to accept the reality that everyone is a part of the sales process. (Yes, even you IT prima donnas.)

Some people are flustered by these things. I've literally seen people devolve from productive and successful to pained and stagnant due to rapid change. I've learned to thrive on change, embrace change, and, dare I say, I even need change. My wife would probably agree that I show all the signs of addiction to start-up culture and she'd be right.

In 2007 I jumped out on my own. After a major course correction from iteration alpha (*we listened to what the market wanted not what we THOUGHT they needed*), my company became the Interactive Marketing firm create-WOWmedia (www.createWOWmedia.com).

But throughout my company's growth, *I never hired a single person.* I made a conscious decision when I went out on my own that I wanted to avoid hiring at all costs, keep my infrastructure and overhead minimal, and run the most "virtual" organization I could — and I succeeded.

As the concept for this book was being assembled in my brain in mid-2010, I started wondering what was next. I craved the ability to focus on one company rather than the deal-flow cycle in a small consulting practice. So I began the process of finding an acquirer.

I wanted to write this book and actually *live* the position I was going to write about. In July 2010, I found a new home with a company right here in Iowa (BirdDog-Jobs.com) who was absolutely looking for my skill set and who was perfectly comfortable with my lack of Photoshop and Design skills. (More on that soon.)

I also sold my company to one of the most amazing vendor/partners that I've worked with over the last couple years. Using the term "vendor" doesn't really indicate how important Andrew Clark a.k.a. "The Brand Chef" (www.twitter.com/thebrandchef) was and is to my business.

Andrew is now thriving under the brand createWOWmarketing (www.createWOWmarketing.com).   Small name change, but new energy, new offerings, and he's rockin' it here in the Heartland.

You may ask : Why write a book on hiring marketers when you've never done it?

In essence, every time createWOWmedia landed a deal it was hired as an "Adjunct Marketing Department" to execute many pieces that involved many sub-pieces, all of which require specialized talent. After all, not every marketer or marketing department across the land can do everything right - right?

I was acting as a a MultiThread Marketer for my clients.
Webopedia (1) defines "Multithreading" as:

The ability of an operating system to execute different parts of a program, called threads, *simultaneously. The programmer must carefully design the program in such a way that all the threads can run at the same time without interfering with each other.*

The operating system is **your brain**. The programmer? **You**.

The MultiThread Marketer must be able to execute many different parts, pieces, processes, and people in parallel while being cautious not to interfere with each other or degrade quality along the way. CPU's do this to the tune of 520,833 MIPS, the **M** stands for **Million** as in **Million Instructions Per Second**. (2)

If your company hires a strategic or **MultiThread Marketer** with a modern, robust, and just-technical-enough skill set, it will make success a far more likely outcome. The right marketing hire will enable your organization to marshal all the best marketing resources it needs to succeed in our modern, turbulent, and light speed economy.

Make the wrong hire and your company will miss market opportunities, be outpaced by the competition, or worse. The MultiThread Marketer doesn't merely fit the Jack-of-all-trades moniker. Rather he has a blend of long-term linear thinking to balance strategy and vision with hyperactive "Googlish" thinking that filters, processes, and executes at ludicrous speed.

If you already have Marketing leadership in place telling you that your company, people, and brand just aren't ready for today's interactive transparency, or that blogs and Twitter are just hotbeds for negative comments, you have a *much* bigger issue to tackle. You've got a marketer on board who's outdated and afraid to expose his weaknesses, to your company's detriment. Your marketer has decided to take the blue pill versus the red pill. If this is your company, prepare to be uncomfortable.

You may ask : Can't I find all of this information on Google without reading your book?

I scoured the web for all of the amazing free information out there on making the right strategic marketing hire. Much to my disappointment, I found mostly the same "How to's" on the hiring process and superficial looks at skills and abilities. I also discovered that no one has truly defined the modern marketing skill set. *From this void... The MultiThread Marketer book was born.*

So let's get started and find out how to identify your next MultiThread Marketing hire to propel your firm to new levels of awesome.

Introduction

Welcome aboard! I hope you're reading this book **before** making your strategic marketing hire. Why? Because in today's economy and business environment, a key marketing hire can be the catalyst to new levels of success, or the blindfold that enables the escape of your greatest marketing opportunities.

If you've already made a hire, I suggest you buy a copy of this book for him or her to do some self-reflection and evaluation. By identifying strengths and weaknesses, the marketer, along with her management, can fill in the gaps and make progress.

Marketing Defined By Me

Marketing as a profession is wildly misunderstood and mislabeled. Most employees, when queried on their definitions of marketing or marketers in their companies, would say things like:

- That's the department that makes our brochures.

- They're the ones who do our graphic design.
- The Marketing Director decides what our brand message is and her marketing managers develop the collateral around that message.
- That's the group that organizes and manages our trade shows.
- They do our website, but it's always a year behind.
- Well I'll tell you, the marketing department is one lucky group because they plan all the parties, the golf tournaments, and corporate events - man what a cushy job.
- They get in the way of making sales because they don't give us what we need to do our job.
- They sell our product.
- A recent Dilbert cartoon described marketing as "Liquor and Guesswork".

I've heard just about all of these from companies of all sizes. How would you describe your marketing person, group, or department?

I define marketing as follows:

Marketing is generating and capturing demand by engaging sweet spot customers and adapting to remain relevant forever.

This process either results directly in a sale or provides access to a sweet spot customer so a sales person may close the business. If marketing does a great job, there's a natural transition to a sale or engagement. Sometimes a sale is a direct result of excellent marketing with no intervention or assistance from a sales person or process.

You probably purchased this book based on the excellent online marketing and infrastructure provided by my publisher FastPencil, my marketing via the web, hearing me speak at an event or conference, or reviews by previous readers. Either way I didn't hand off a lead to a sales team to close your business. You bought based on my marketing.

Don't Know Everything - Just Know Where to Find Answers

I'm not sure at what point in our careers or in company life we become threatened by the falsehood that everyone must know everything to be valuable and valued, but it's reality. I see this most in the marketing profession and it's crippling. Be comfortable with what you don't do well and either learn the skill (if that makes sense) or use an external resource who has it.

I built my small company on the premise that the sum of its amazingly skilled non-employee "parts" creates WOW! well beyond what individual pieces can achieve. Take pride in not knowing everything because it demonstrates maturity of thinking and self-realization/actualization. Do you believe that the CEO of Company X, Inc. thinks she knows it all? If she does, then why does she have a CMO, CTO, COO, and myriad other acronyms to help her lead and run the business?

Increasing Velocity of Disruption

Market velocity is accelerating and the very definition of marketing is being reshaped as I type these words. Topics like search engine optimization (SEO), keyword targeting, online reputation management, and content outposts have entered the realm of the marketer, no longer reserved for propeller heads in the IT department. It's impossible to separate online and offline marketing functions into distinct chunks because the functions are so intertwined.

For example: Company X engages someone on Facebook and provides a printed coupon that drives the customer into a store where she signs up for e-mail updates, a paper catalog, and an in-person follow up appointment. She may provide some feedback on Twitter that immediately reaches 10,000 other hyper-connected "digi-moms" in the Twittersphere:

"Had an amazing experience at Company X store. Go see Jane. Amazing #productX prices. Coupon code XYZ for 25% off one item".

Just yesterday I read about an amazing customer service experience that a prominent blogger had with Adagio Teas (www.adagioteas.com). I was moved by the experience, the product, and its simplicity. In about 4 minutes I had:

- Read about a company I'd never heard about before.
- Followed them on Twitter (www.twitter.com/adagio-teas [http://www.twitter.com/adagioteas]).
- Received a $5 off coupon code back from the company in a Twitter Direct Message (DM).

- Placed an order for their starter set at $19 and purchased even more sampler packs to reach their $50 order free shipping threshold.
- Tweeted about how impressed I was with their firm.
- "Liked" them on their Facebook Fan Page.

This is of course more likely in a B2C environment where a passionate impulse buy is possible, but this kind of passion, conversion, and organic promotion is absolutely possible in B2B environments too.

We no longer have the ability to escape the wrath of a disappointed customer nor can we decide absolutely what our brands will become in the market place. We can provide fuel to power the marketing engine but the steering wheel has the hands of each and every one of our customers (and potential customers) on it. Don't fall asleep at the wheel.

The Incredible Shrinking IT Department

Do you even have an IT Department at your firm? The small business economic engine in this country is increasingly being driven by hosted or "SaaS" (Software as a Service) applications removing the need for expensive licenses and hardware. I ran my company on about $200/month worth of hosted software including accounting (Quickbooks Online), specialized invoicing software (Freshbooks), project management/collaboration software (Central Desktop/Basecamp), content hosting (Amazon S3/Dropbox), blogging and company website (Wordpress hosted at MediaTemple), and I'm even writing this book using the amazing hosted publishing solution Fast-Pencil.com (Not a paid endorsement. I wish.)

So instead of being held hostage by IT, most small and medium sized companies can spend a much smaller monthly amount per person or per user and bypass barriers like hardware and infrastructure management that held them back in the past. Thus we can test technology-based marketing and campaign-management tools without even asking permission.

What used to be an "enterprise roll-out" and typically a very expensive and painful plunge into over-promise and under-deliver is now a cheaper, safer playground. For more stories and insight into the Enterprise Sales process, you can read my free ebook, "Confessions of an Ex-Enterprise Salesperson: What I Really Meant When I Said_____" (Available at www.douglasemitchell.com)

I Know Half of My Marketing Works - I Just Don't Know Which Half

Marketing metrics are very easy to gather today, yet most businesses we encounter don't know how to analyze and make decisions based on the data. If you're a small-to-medium sized business (SMB) you must have marketing leaders who can make sense of the numbers and who know when a deep dive into more complex analytics is needed. My createWOWmedia team leveraged metrics to show the effects of our actions. We also used data to help our clients spend far less on marketing activities that yielded little return on investment (much to the disgust of on and off-line media salespeople). Your marketing leadership must have a bit of 6 Sigma statistical nerdery in it to win this game.

Agile Candidates for Agile Companies

For the purposes of this book, we're going to look at hiring marketers for agile companies: Small and medium sized enterprises with few layers of management and bureaucracy.

Wonder which one you are? Ask yourself:

"Can someone propose a change in our business and within a very short period of time a decision is made, the proposal is funded, and execution is begun?

If committees and endless meetings are required to do things at your firm, and if thwarting progress via policy and management road blocks is the norm, then please read on — but don't tell your boss.

Large companies are welcome to read this book and apply its thinking to their practices. But large company hiring is often so far removed from reality that my approach may cause temporal explosions (think about the movie "Scanners").

Maybe someday I'll write, "How to Make a Marketing Hire for the Massive Enterprise" that will include a case of beer, some photos from HotorNot.com, and a dartboard. But for now, we'll stay distinctly in the realm of agile companies.

As a framework for each chapter, you will navigate:

1. A personal story or anecdote of my own.
2. Creating your own personal story.
3. A lesson from a third party story.

Now Let's Break Down The MultiThread Marketer Into Component Parts.

CHAPTER 1

Section One: Finding Your MultiThread Marketer

The MultiThread Marketer's skill set is a moving target, but we'll take a snapshot of it in the pages that follow.

No position has transformed more rapidly and left more in its destructive wake than that of the Marketer. In fact the skill set has changed so quickly that most companies don't even know they're outdated. Sure, sales may be steady and profits may be hanging in there for some, *for now*, but this will change without the proper talent in place to drive through the tumult. Your firm must hire right to catch up and have an excellent guide through the modern marketing battlefield.

The new marketing leader is adept at sniffing out trends, skilled in web research and business intelligence gathering, and comfortable with online collaboration tools. He is capable of project management for both internal and external resources, able to communicate in all forms including audio, video, visual, and written, has skills in leveraging social and interactive tools to build audiences, and he must know enough about technology to set expectations and manage results.

The MultiThread Marketer is a strategic ring leader of sorts. While he need not be an expert in all areas, exposure to these key areas, skills, and processing, and working through them with partners is critical.

Let's do a deeper dive now and deconstruct our Multi-Thread Marketer.

CHAPTER 2

The Basics: Seriously?

The Two Most Valuable Classes I've Ever Taken

Twenty years post-matriculation from my alma mater Glendora High School, I still strongly believe that the most valuable classes I've ever had (in high school or in all collegiate work combined) are typing and speech/debate. Typing class gave me the power to sit down and let thought flow through my fingers - fast with no fear of what was flowing out. That's what editing is for. Speech taught me how to see both sides of an argument (since we were often asked to take the side of an issue we explicitly disagreed with). During my career I've seen deeper into issues and peeled away layers when company leaders had very strong opinions about strategy or execution, and I saw a different path I wanted brought to light.

Peck Peck Peck

What I'm about to say would sound ridiculous and sophomoric if it weren't true. In my experience working with marketing clients and departments, about 50% of those over 40 years old cannot touch type. Over 50, about 80% cannot. I'm making statements based on my experience - not painting age brackets with summary judgment.

Marketers who can touch type 60 words per minute are far more efficient at creating text-based content and communication than someone who hunts and pecks on a keyboard. We're not even talking about the quality of the words that come out yet.

I've seen VPs of Marketing (and CEOs of companies) take 3 minutes to construct a sentence or two for an e-mail as I sat in amazement. As I said to start this chapter, the most valuable classes I've had since high school are typing and speech.

If you're hiring, you must ask your candidate straight up, "How many words per minute do you type?" Perhaps you formally test them with some kind of free web tool. I don't care. Your candidate MUST know how to touch type and be pretty quick about it. And if YOU can't type, please seek help. You don't need a semester and a classroom anymore either.

Put your candidate in a situation where they need to create something for you. Watch how your candidate interacts with a keyboard and mouse. Does he know any shortcuts to save time? Does he know where certain menu items are in PowerPoint (cringe) or Word if you're a Microsoft house? I'd rather have my VP of Marketing be focused on creating an amazing visual and interactive chat with a potential partner or client than have them banging away trying to figure out how to make the company logo fly in. You think I'm kidding?

Must Play Well With Others

Get a very clear sense of how your candidate interacts with others and how he has interacted with others in the past. Is he comfortable managing projects and teams? Is he good with delegation yet still able to take ultimate responsibility for a project? Listen for keywords and phrases during your interviews like:

- We
- Us
- My team
- I was responsible for
- Autonomy
- Giving authority
- Leveraged external resources
- I outsourced that because it's not my strength
- I managed multiple projects simultaneously

Recently, I pitched a company some ideas on driving demand through content creation online, i.e., inbound marketing. Just before the meeting I was told that their Marketing Director had been "asked to step down because he delegated quite well but didn't actually do much and took no responsibility for the delegated results". Not good.

As an agile company you don't have the budget to simply delegate everything, sit back and take credit for the good work anyway, but continue to probe for ownership and teamwork.

This book isn't about psychologically profiling candidates, but let your gut be your guide. Your hire must type fast and play well with others.

MultiThreading Foundations

Hunt-and-peck typing was far more commonplace through at least the 80s until word processors began their rise. As availability of home computers and word processing software increased, so did the desire to become more typing efficient.

Entering the 90s, the advent of AOL & Compuserve which included chat rooms and message boards pushed more users to learn touch typing to keep up with their strings of conversations online while more and more every day work tasks were making their way into the desktop PC versus happening on paper.

The legendary **Mavis Beacon Teaches Typing** program, released first in 1987, is *still* available today to guide the hunter and pecker to a state of typing zen. In case you were wondering, "Mavis Beacon" is not a real person; the original photo of Mavis Beacon was of a retired Carribean-born fashion model named Renee L'Esperance."(1)

CHAPTER 3

The Search For Knowledge

There is NO WAY I'm buying a new toner!

So my HP printer throws an error message telling me to replace the black toner. I try and try and try to get rid of the message to no avail. "I know there's more toner in this damn thing! Next stop, the Google search box: "HP 2605DN Toner Error". Results: within a few moments I'd found a fairly elaborate workaround that involved pressing keys on the printer in combinations, entering "service mode", and overriding the message allowing me to print for about another year before buying a new RED toner. A couple minutes of searching saved me nearly $100 and gave me a sense of satisfaction that I'd beaten "the man".

How Do You Google?

Just knowing how to search is a valuable asset in the MultiThreader's toolkit. I can't count the number of times I've asked someone to "Do some research on the web and build me a picture of what's out there" only to be told that "There's not much out there, really". Really? Really?

Searching, digging, and plodding through the archives of the universe online can be daunting. Since this book is not a comprehensive "How-To," I'll leave it up to you to discover the methods and search strings that can make a huge difference between getting crap results and hitting the jackpot.

I've barely begun to scratch the surface here as my work typically involves fairly mainstream topics. One of my excellent freelance bench members is very adept at delivering "Business Intelligence" - hopefully not an oxymoron in your firm. He can dig up primary research and find reports that build a much clearer picture about people, markets, and plans. I call this kind of work "Business Plan Killers" because it typically exposes the hidden facts that dismantle your business (new or existing) assumptions before your eyes. Don't get me wrong here, most businesses are started from a passionate fire in the gut and in a meaningful percentage, persistence beats data.

For a nifty cheat sheet on getting the results you need from Google, I like this reference from GoolgeGuide.com (http://www.googleguide.com/advanced_operators_reference.html)

Search.Twitter.Com

Searching Twitter for knowledge and trends has become a very powerful tool in the multithreading arsenal. By visiting http://search.twitter.com you can get a "real-time pulse" on topics. This research can be extremely valuable in deciding how to micro-target certain audiences with long-tail keyword combinations, i.e., three and four word keyword combinations that get far fewer total numbers in search results, yet get very high click through and depth due to their specificity.

MultiThread Search Lessons from Monte Python

Bridgekeeper: What… is your quest?
King Arthur: To seek the Holy Grail.
Bridgekeeper: What… is the air-speed velocity of an unladen swallow?
King Arthur: What do you mean? An African or European swallow?
Bridgekeeper: Huh? I… I don't know that.
[he is thrown over]
Bridgekeeper: Auuuuuuuugh.
Sir Bedevere: How do you know so much about swallows?
King Arthur: Well, you have to know these things when you're a king, you know.

King Arthur was "The Google" of his time.

CHAPTER 4

How's the Candidate's Google Juice?

Being findable online is a conscious decision.

Say that again out loud. If you are lucky enough to be the only guy named "Joblugahbutz Jasperwitz" congratulations and welcome to the first page of Google results. If you're like the rest of us, you need to work at being findable when people type your name into the magic search box.

I received a call one day from a political consulting group from another state. The gentleman asked me to do a clandestine focus group video with a group of possible candidates for a race here in Iowa. I was paid well and during our first call I asked how he'd found me. His answer, "I searched Google for 'Video and Multimedia in Des Moines' and you were in the top results. While he did not search for my name specifically, he did search for keywords that I'd consciously chosen to put into my Google business listing,

*my company profile, my blog keywords, etc.   I was found
because I chose to be found.*

Where Do You Look For Stuff First?

Where's the first place you go to gather information on
people, products, places, or services? (If you're like 99% of
my audiences over the last 3 years, you said "Google").

Will you head online to check out your new candidate's
background? You'd better. In fact, "45% of employers
reported in a recent CareerBuilder survey that they use
social networking sites to research job candidates, a big
jump from 22% last year. Another 11% plan to start using
social networking sites for screening."(1) I'd expect these
numbers to be much higher still, especially within SMBs
who typically lack more robust HR data gathering tools.

So how does your candidate fare when you search their
name? Is she findable? I'm not suggesting that your candidate be on the first page of results for their name alone
(especially if it's common like Joe Smith). However I am
suggesting that there should be some kind of **Personal
Brand** out there for the marketer you think you'd like to
hire.

If you're interested, search "Doug Mitchell" and see what
you find. Choose the different types of specific results too
like "image" and "video". I still haven't overtaken the
Paris Hilton song singing Doug Mitchell (do the search
and you'll see) but alas, what's more important in this
world?

By having more Google juice, your candidate is demonstrating that he is in the game, putting himself out there, and establishing a personal brand. He's also demonstrating that he has brand equity to bring to your firm. Danger, Will Robinson: If your candidate isn't findable on Google - at least within the first two pages - there is a problem. If they are unable to market themselves as a brand, how are they going to succeed at marketing your brand?

If you can watch 30 videos of me talking about how to do amazing things with your company's marketing you'll have a pretty good idea of what you'll get when I come and talk to you. When you read the case study createWOW-media released in 2010 (2) you'll know what I have delivered. If you listen to any one of the 30+ Internet Business Podcasts I've done (www.ManagingtheEdge.com) you'll know EXACTLY how I feel about making technology your servant, the power of outsourcing, and the responsibility of companies to create compelling content to earn page rank.

I've often had people tell me that they feel like they know me already because I talk and act in person just like I write and act in videos. Having Google Juice simply eliminates many of the hiring questions. Will she be a leader for our company and interact with the press? Will she write for our firm or will content flow be minimal because she's scared to make a mistake in public? Will every customer testimonial be a $5k production with a high-end firm or can she grab a camera, shoot some footage, and produce it into a powerful piece in a day's work?

Like the academic community, marketers are more respected and have more authenticity if they publish, i.e., put themselves out there for criticism.   If you find a treasure trove of information, writing, tweets, videos, podcasts, etc., on your future hire you'll be MUCH more secure in your decision.

Don't Be Afraid. Embrace the Rockstar Brand!

In countless companies and agencies, I've seen talent walk and talent be summarily rejected before the interview stage because *the potential hire's personal brand is BIGGER and MORE ROBUST* than the brand of the company doing the hiring.   I call this the "Rockstar Effect". Companies typically don't like it when their people are more popular, findable, and potentially valuable than they are.

While the odds of a rockstar marketer having more opportunities to leave and do other things are great, you should do everything you can to find them. Rockstars bring a following and if you are the right fit (not every rockstar is the right hire for your company, mind you), the rockstar will lift your brand and bring his or her mojo to you. The rockstar hire may force your company to be uncomfortable and to stay on its toes as new doors open and new PR opportunities abound. Get used to it!   Embrace your inner Rockstar.

Under no circumstances should you put the lock down on your new hire's brand in the hopes that you won't be overshadowed.   You NEED that rockstar to be out there speaking, writing, and being visible. Your firm will reap the benefits over the long haul.

It is a good idea to have a policy discussion about self-promotion during work hours to get that on the table. If the rockstar argues that writing five blogs and tweeting 20 times per day on her personal account will benefit you, she needs to clearly lay out HOW. The goal in paying said rockstar is to transfer the uber following and awesomeness to YOUR company. So, unless the rockstar is merely a paid representative or mouthpiece, do have the discussion.

It's really easy to gauge volume of output so if you see four posts in a month on your company blog and 20 on the rockstar's personal blog, you may need to have an uncomfortable conversation.

Marketing Lessons from Headliners and Rockstars

Think about how sports teams try to sign or trade for big name players. Yes, they get a lot for what these superstars do on the field, but there are other benefits. Superstars sell more tickets. They end up doing commercials and ads while wearing the team uniform. And they appear in all-star games, on talk shows, and on magazine covers. Yes, a lot of these are individual efforts and accolades, but at the same time, they also are building the brand of the team that employs them.

CHAPTER 5

Measure Twice - Cut Once - 1300 Times

1300 is the average number of searches for this chapter's title globally each month.

Practice What You Preach

As I poured over the analytics and lamented my poor performing site to my web ninja , I came to a stark realization. "I'm writing content and and focusing on keywords that will attract an audience that will never buy this online information product." It was one of those duh moments that feels like a breakthrough but is tinged with a little WTH on the back end. He said, "Yeah, happens to the best of us but at least you caught it now." "I can't believe I did what I specifically tell my clients not to do," I shouted.

Then we rolled our sleeves, made required changes based on research and data, and turned things around within weeks.

I was achieving the goals I set after getting crystal clear on how ruthless the web is. Data based decision-making rules in most online projects where we desire real results, conversions, and dollar bills.

The Web is a Ruthless Number Cruncher

Web analytics are ruthless. Google has become massive by understanding behavior through search. The Googlebot knows "How we think" or at least currently has the best simile of how we behave online. We can sit down together and find out exactly how many people searched for "Ultimate Steak Seasoning" or "Ultimate Prime Rib Recipe" and determine if anyone is bidding on those key word combinations and what it would cost to "buy a click" if we placed an ad. Your early and eager embrace of these facts will help you remove the silly emotion out of your tactical decisions. Note: I don't suggest removing *passion* from your decisions. Just be aware that no matter how much you believe you have the next pet rock, you are invisible if no one is searching for that term. (In case you're wondering, as of 6:10am on January 28, 2011, there were 40,500 global searches each month for the term "pet rock". Happy selling.)

The Analytics Concepts Multithreaders Need To Care About

This book is not a web analytics primer (and for this chapter we'll specifically be talking about the free Google Analytics product). Frankly, your own online searches will yield vast resources online that answer 99% of your

analytics questions. However there are some key perform-
ance indicators (KPIs) that you should have a working
knowledge of as a MultiThread Marketing leader.

Bounce Rate: You want this low. This means someone
found your site through a search or paid ad, clicked to it,
and said, "Oh this is so not what I wanted", then left
without clicking anything else. Bounce rate is a quick and
dirty indicator of relevance as it relates to how people
found you.

Unique Visitors: Daily unique visitors are a metric worth
measuring. I can't tell you how many times I've seen
someone's office manager use the company site as his
browser's "home page" to help the traffic numbers.
Actually this skews the numbers. So we're looking at
Unique Visitors as a valuable measurement.

Traffic Sources: You'll want to understand where your
visitors are coming from. Notice the pie chart on the analytics dashboard for your site once you have Google Analytics up and running (about 2 minutes if you use Word-
Press and an analytics plugin). You'll see the breakdown
of direct traffic (typed in), search traffic (obvious), and
referral traffic (someone clicked to your site based on a
link elsewhere). Watching this dynamic change over time
is fascinating.

There is no perfect ratio as each site has different goals
and traffic sources. If you're a pure content marketing
player, i.e., you're blogging to attract organic visitors
through search, then you'll of course want a high per-

centage of traffic to arrive via organic search. You get the idea. Some sites are purely driven by Pay Per Click Advertising (which I'll discuss in more depth later) and could have low organic search numbers and high bounce rates and still be successful.

There are more analytics concepts to be aware of but please, do the research. Google provides plenty of education on it and there are millions of blog posts that will teach you everything you need to know when your analytics ninja gives you the weekly report or you check in on your dashboard.

Using Analytics to Test and Launch Businesses

One of the most valuable and least utilized ways to leverage analytics is to test business concepts and launch "mini-projects" that could become full fledged online businesses. In a few hours, even mere mortals like me can put up a site or landing page with an e-mail form with A/B testing based on some keyword research and start collecting results. If I have a larger business idea in mind, often I can save myself the trouble, money, pain, disruption, and opportunity cost of tackling it by testing the concept online.

Consider my methodology:

1. Do keyword research and exploration for 4 hours.
2. Set up a landing page or pages, a WordPress site with interest gathering tools.
3. Talk about it online, drive traffic organically, and use social tools to increase exposure and awareness.

4. Refine the message and talk to people in your sphere of influence. Revise. Revise. Revise.
5. Make a decision about moving forward based on actual demand and feedback.

Total time and money investment (assuming you have no web ninja skills) = about $1000 and perhaps 40 woman hours. Ninety-five percent of the time, this model will tell you this idea won't fly. The 5% is where you turn it on.

The way typical classes and government funded "How to Start a Business" courses do it:

1. Develop your idea.
2. Do market research on your idea to see if it will fly (perfect opportunity for web testing here for your "T-shirt business"). Use lots of government data like "Of the 6.8 billion people on the planet, some 2 billion buy at least one t-shirt each year. If we just get a half of 1% of the market"…you get the point.
3. Write an elaborate business plan full of wild-eyed and unproven assumptions.
4. Continue to believe there is bank funding somewhere out there if you just pitch enough banks that your start-up has legs.
5. Search for government grants and other sources of cash.
6. Build prototypes or a first run so you have a demo model.
7. Attend a trade show or local exhibition show with your idea for research.

Total time and investment is impossible here. There are simply too many variables but frankly, I have personally seen this process take 6-12 months and over $10,000 to reach the "There's no there, there" conclusion.

My MultiThreading Counterpoint

I will tell you plainly that many of my "launched concepts" have not been fully vetted by emotionless statistical analysis. I'm full of passion and not suggesting that you can't make your idea work with brute force. I'm providing what I believe to be the best chance of success.

As a MultiThreader, I am exposed to massive volumes of products, projects, ideas, new landing pages, businesses started over a weekend, videos, concepts, false starts, books, blog posts, tweets, and marketing messages every day. I have a different outlook on many ideas because I keep up with many "threads" out there in my local market and online. You will start to develop a similar view of the Matrix after your immersion into multithreading.

For an excellent deeper dive on using data to drive decisions, check out this episode of the Managing the Edge Internet Business Podcast I do with my web ninja Andy. (http://managingtheedge.com/episodes/data-driven-strategy-episode-26)

Lessons in Analytics from the 2008 Obama Campaign

Dan Siroker quit working for Google in 2008 and went to work for the Obama Campaign. Siroker built a war room full of brains that relentlessly tested messages, images, and content over and over again.

Siroker shares these analytics lessons with us:

- A candidate (or company) *must* define success in a measurable way.
- We must question assumptions.
- Divide and conquer.
- Don't re-invent the wheel.
- Take advantage of circumstances.

Siroker helped drive the Obama campaign to raise *over a half a billion dollars online.* (1)

CHAPTER 6

Mentors: The Fresh Makers

"I'll never turn down a sale!"

Those immortal words still ring in my ears. It was 2003 and after a few pitch sessions to venture capital firms, we'd gotten interest. We had a team of investors doing due diligence. We were getting coached on how to truly hone in on our target market and to stop making the naive mistakes of nascent technology entrepreneurs. I was eating up EVERY minute of interaction with this group of wealthy, successful, and tested entrepreneurs.

"We see a total market of 10.5 million widget buyers and if we just get 2% we'll make a grip of cash" *the CEO explained.*

The investor team pressed harder. **" We're not interested in such a wide target market. I'd rather see you focus in**

on 100 customers in a niche, land 30 of them, then leverage that dominance into other niches ".

Then the CEO floated this little nugget:

"Hey I don't care what the market is or what size the company is! If someone comes to me and wants to buy my product - I'm givin' it to 'em!!!"

Then silence.

I think my leg twitched under the table, like it wanted to reach out and inflict some kind of damage. @!#$%& were the only words that came to mind. After 20 seconds of shock, awe, and bile raising silence, I stammered out something only Tommy Boy would be proud of.

The meeting adjourned shortly after that. There were no more meetings. There would be no $1.2 million or coaching from seasoned tech veterans. This angel round lost its wings before my very eyes.

I will not accept this

Within 24 hours and a few beers, I reached out to the lead investor and laid it out.

"I know why you didn't fund our company. But since we get along quite well and I respect what you've done in your life and career immensely, would you consider mentoring me so I don't make the same mistakes again and let me model your actions so I can someday become an angel investor and help start-up companies succeed?"

We determined a meeting schedule and I was asked to pre-pare a "Who is Doug?" summary. That relationship still continues today.

The MultiThreader's Perspective: Find a mentor NOW

I'd say I've been blessed to have amazing mentors in my life, and that's an accurate statement. **But damn it! These people didn't fall into my lap and they won't fall into yours either.** I *identified* them and I *asked* for the relationship. I guess I AM a closer, Mr. Baldwin. Do I get the Glengarry Glen Ross leads now? (5)

I'm sure mentor relationships do evolve naturally, and in larger companies they are created by fiat, but largely these relationships are cultivated by proactive and dynamic leaders who leverage the knowledge, systems, and approaches mastered by others to compress the learning curve. Tony Robbins talks extensively about this in his books. Some call it "modeling". A mentor shows you the ropes because they've been there and done that and THAT saves time and wasted effort. You MUST seek mentors out and close the deal.

As a MultiThreader you must conserve processing power and be careful not to degrade the quality of your threads.

Mentors will help you achieve this goal.

Mentoring is not simply a "gimme" one-way street either, so understand that these relationships aren't developed purely as a means to your success. My mentoring rela-tionships have perhaps started out tilted in the ME direction. Over time, however, as my skill, experience, and network grew, my ability to give back to the mentor grew. Some of those relationships have evolved into closer friendships with less formality and business talk.

You need to act as a mentor as well. If you find someone professionally or personally that you connect with, who has a spark that you recognize from your own persona, make yourself available. Give back. Be a mentor.

MultiThread Lessons from NASCAR

Today most star-potential drivers enter top divisions of my favorite sport near their eighteenth birthday. These kids have been racing all their lives and have probably come up through a "Young Guns" program with a top racing shop.

During their years in those programs, the drivers are carefully guided in tire management, crew communication, race strategy, pit strategy, and car set up, media relations, and business management. When the kids enter NASCAR's top series, they've got the complete package with a few rough edges.

This "multi-mentor" approach produces accelerated results that benefit all parties. Consider yourself a burgeoning multithreader with star potential no matter your age. Build a team around you, and practice your victory speech.

CHAPTER 7

Weed Whacking

My Three Miles to Business Zen

For the last year or so I've been walking with my good friend, mentor, and colleague Michael C. Wagner (www.michaelcwagner.com). During our walks he shares his wisdom with me. Mike guides companies to work on their businesses instead of in them. That may sound cliche' by now but no greater challenge faces businesses in today's hyper-economy. There's simply too much to do, too much to process, and too much information coming in.

Mike guides leadership teams (and me) through his lessons, learnings, and sharings to stay out of the weeds so decisions can be made across an enterprise without requiring constant leadership presence. If a company has a decision-making context that's in alignment with its goals, culture, etc., then leadership can lead and companies will thrive. My walks with Mike consist of 60 minutes of straight head-clearing banter that send me back to the office with a renewed sense of purpose. I'd highly recommend that you find something or someone that frequently brings you back into business zen because before you know it a quarter, a year, and a career may have slipped by.

Fireman vs. Fire Starter

How often do you feel as though you're an underpaid, under-trained, corporate fireman? How often do you feel that you're in complete command of your company's marketing destiny, lighting fires of awesomeness throughout the company?

Most will say they feel out of control and at the whim of the market, their employees, their external resources, their management, etc. This concept of constantly being in the trenches, putting out fires, running around like mad, and reacting to circumstances, instead of leading, is what I call "Getting caught up in the weeds."

The MultiThreader knows the weeds are something that cannot always be avoided. Look, the little buggers can grow almost anywhere. They have amazing resiliency, and despite our best efforts, they cannot be eradicated. Yet, the weeds can be kept under control and not left to get too high. When marketers are running around in the weeds they cannot see the big picture or what others are doing.

The MultiThreader knows that with the proper use of pesticides, and by simply removing the weeds when they're small, her vision will not be obscured. Keeping the weeds down means she can survey the entire landscape and notice issues, good and bad, throughout the marketing landscape. Sometimes we let the weeds get really high and we have to spend some money to get them cleared away.

This "gardening" practice is not easy. In fact since business rarely operates in a perfect timeline immune to outside influence, the MultiThreader must shrug off the myriad opportunities for derailment, and when that shiny object appears, she must rapidly compare it against the strategic plan, long-term goals, and vision for the firm. She must also realize this approach isn't perfect. However, being cognizant and aware are key.

Sometimes when I'm in an execution phase or have gone down a necessary rabbit hole and colleagues ask how I've been, my response is "In the trenches pruning some weeds." This isn't derogatory mind you. Sometimes we must see things at ground level or perhaps even get a little subterranean to have a better perspective again when we're looking at things strategically and leveraging our resources to execute MultiThread Marketing.

Strategic Plans Are Your Friends

I know often we're asked to create marketing plans for a quarter or a full year by our executive management or boss. This strategic planning process can be disheartening especially if you're in a company or a department (perhaps yours?) overrun by weeds. The best laid plans are often scrapped at the first sign of greenery.

Leverage your strategic plan frequently if for no other reason than to spark an assessment of your "weed level". There are enough distractions in the modern workplace to derail anyone and obscure reality. "I'm just so busy" is something I hear from 95% of business people I interact with. Don't succumb to "busy-ness" because it's a surefire way to become mired in the trenches.

MultiThread Marketing Lessons from Big

In the movie, Big, Tom Hanks gets caught up in "the weeds" of busyness and forgets who he is (he's still a kid trapped in an adult body). While he might be showing signs of success, he is losing the "edge" that brought him the success in the first place - his child-like perspective. It takes a heartbreaking awakening to re-realize who he is and what he is about.

CHAPTER 8

Whatcha Been Readin' Lately?

I'm Starting to See the Matrix

I have a particular passion for this chapter because of my own experience over the last three years. Have you gotten so busy that you aren't able to pick up a book, listen to one in the car, or read one on your smartphone or iPad? I was that guy too. But my passion for reading and listening was reignited and a funny thing happened.

I began to see things in business and life in a much broader context. I leveraged the information I was consuming in book after book to build a "support network" for my decisions based on well-thought-out and well-researched material. In other words, like Neo in The Matrix, I began to make sense of all the ones and zeros streaming by. I saw patterns, behaviors, and tactics that I'd read about over and over again. I'm better able to postulate about outcomes and results because "I've been there before." Passionately con-

*suming information from smart people is keeping me deep
in a shallow world. (Thanks @bigwags)*

Stay Deep

Small-company life had taken all of my mental band-
width. The modern and "shallow" thinking that Google
encourages and Nick Carr so eloquently writes about, (1)
had stolen my ability to focus between cell phone beeps
and IMs. When life starts to seem like 140 character
vignettes, it can be challenging to read between the lines of
interpersonal communication between co-workers, contractors, colleagues, and your spouse.

But somewhere during the last six months, and with the
help of key mentors, I re-initiated the launch sequence on
knowledge acquisition through reading and listening to
audio books. This decision rekindled the fire and desire to
gain depth through others. I'm still far from prolific but
I've cranked off about 10 books in the last three months.
That's 10 more than the past two years frankly.

Many of you will say, "I love reading fiction! But I just
can't read in my industry." I'm the opposite. When I ded-
icate time to reading, I'm not trying to relax. Rather I'm
trying to fill my mind with ideas for distillation and inte-
gration into my daily life. I'm starting to notice patterns
and trends in reading some of the best books of the day. I
can see how threads of research and principles start to
show up across great books with completely different big
ideas. This is really exciting to me and it's making me a
better marketer and leader.

If you're hiring a marketing leader or assessing your current marketing staff, you need to become curious about their consumption of information. You can glean valuable information with a simple, "Hey have you read the book Drive by Daniel Pink? There's some amazing stuff on what motivates people in life and business." Pay close attention to the response.

Your marketing leader or candidate should be able to riff off a few key relevant modern books in the field, and perhaps a few more that have stood the test of time in business. This line of questioning isn't designed to imply that if your marketer doesn't read, he isn't qualified, but it does give you some perspective on, well - his perspective!

My mental picture of the modern MultiThread Marketer (and more importantly perhaps the modern MultiThread Consumer) has been boosted dramatically in the last six months. I've deepened my appreciation for what it means to build a successful company by developing customers first and products second. I'm rounding out my online and offline marketing toolkit to execute successful high-level strategies and executing like a fine tuned military unit.

These new refinements, perspectives, and skills are born of reading and listening. You and your marketers should be doing the same. If you are a current marketing leader, build an education and book line item into your budget. Most are under $20 these days and if you spread the books out over your team, you can rotate them to save money.

Let the best bloggers do your research for you

The reading of books alone will ratchet up your depth-o-meter. But adding in industry blogs, many written by the same book authors, will continue to build your base. Business books are frozen moments in time. They may capture current events or represent emerging trends but blogs are a continuum of discovery.

Most often, the best writers these days tease their book ideas, showcase their new research, and test ideas on their own blogs.   Writers will often expose you to other research, experts, and authors in the field as they build knowledge around their thesis.   Often one blog post will point to three additional articles that, when combined, paint a beautiful picture of depth and clarity. Many have research assistants and staffs. Let them work for you and let their publishers pay the bill.

When the author's next new book is released, you'll be in the loop and primed. Few books on business are written in a vacuum today. For a list of blogs I consume with passion, see the references page.

Stop wishing you had more time and make this happen. The MultiThread Marketer must continually be assimilating new perspectives, theories, and thought threads into his own knowledge base to remain relevant in a world that passes by very quickly.

Take It to the Next Level

One of the most valuable investments I've made in myself is an Audible.com subscription. Audible brings downloadable audio files down in price and provides easy delivery to my iPhone. I use the 10-20 minutes a day spent in the car or 60 minutes over lunch to listen on 2x speed to boost my consumption. (Yes - I moved from California to Des Moines and part of my reward is almost no windshield time and about 7000 miles of driving per year.) When combined with book reading, I can inject 4-6 new books into my working knowledge base each month. Invest in yourself. Fill in travel or down times with enjoyable learning experiences. They will pay you back.

MultiThreading in the Matrix

Once liberated via the red pill, Neo is "trained" by Morpheus' crew by plugging in and having programs loaded into his brain. These programs provide him skills, context, and knowledge from which he pulls constantly when making decisions or fighting against Agents in the Matrix. Reading and consuming audio books is your training for the Matrix. Absorb as much as possible like Neo did, and enjoy the rush!

CHAPTER 9

Section Two: Becoming A MultiThread Marketer

If you've gotten this far, you may be closer to becoming a MultiThread Marketing Rockstar than anyone you're considering hiring for the position.

One of the recurring themes of this section is a three-pronged approach to the many "departments".

Learn just enough to be able to:

1. **Set Expectations**
2. **Manage Results**
3. **Know How Much Stuff Should Cost**

That's a big part of being a MultiThread Marketer. You won't be taken advantage of in terms of either time or money. So maybe, just maybe, you're the candidate you seek. Let's read on and see.

CHAPTER 10

I.T. Depends

The last five years have seen an explosion of web based or hosted applications at affordable prices for most agile businesses. It's almost overwhelming trying to keep up with the latest simplified tool to tackle a specific niche problem.

Early in this book I mentioned that I ran my business on handful of web apps that "in the old days" would have required IT knowledge of servers, etc. To date I've never involved an "IT person" to set up or use ANY of the apps that run my business.

Your ability to work around the IT department and just "Get $hit Done" is unparalleled today. Life as a Multi-Thread Marketer it seems, is good.

But in the last few years I've paid tens of thousands of dollars for programming and IT assistance for client projects. I've also worked with (and navigated through and around at times) IT departments and outsource houses for client projects. In my current position I work with an IT staff (of one) to execute and coordinate our strategic marketing plan where web apps dramatically intersect with "old school" apps and web sites constantly. When I joined the management team of my first start-up, I learned how a five person programmer team can produce amazing results, while simultaneously driving product development down a wormhole.

As a marketer you WILL have to work with, inside, and around IT to accomplish your marketing goals so get comfy with that, take a deep breath, and read on.

Learn Enough Code to Set Expectations, Manage Results, and Know How Much Stuff Should Cost

Each time I've made a breakthrough in managing the IT-Marketing combination...I really did feel like someone sat me down in the chair, plugged my brain into the matrix, and uploaded a "training program".   I remember how desperately I wanted to stay out of the weeds on these things and not get dirty in the trenches.

One day, I'd reached an impasse on making some basic formatting changes to a WordPress website we were developing for a client. I realized I'd reached the end of the road and that no amount of searching would solve my problem.   After wasting hours noodling for the right "widget" or "template setting" for the changes (i.e., the easy route) I tweeted out:

"Anyone in DSM that knows WordPress - Need Help ASAP - Please Help"

Within hours I was working side by side with a local guy I "knew" from the Twittershpere. I paid him hourly to execute as I watched and learned in tandem. The lights went on and the training program burned new concepts into my brain.

Take note: I was paying to have work done and going to class at the same time, not because I wanted his job but because I wanted to be a better marketer.

I'd learned where things were and how fast things could be changed, more than the exact thing to do or code to write. This is a key point. Now I knew how long some key fixes, code changes and adaptations should take, and could have a more realistic expectation setting. I wasn't blindly asking someone for help, sitting back wondering why it takes so long to make a simple change on a web page or to move an image or make a landing page template, etc. I knew roughly how long that would take, and perhaps more importantly, what I should be billed for the task.

Nothing is more frustrating than watching technology-clueless marketers whining about developers not doing what they want because it should be "fast and easy." If you're in IT you're nodding your head and giving me props right about now but hold the gravy. One thing might actually be more frustrating and that's prima donna developers yipping like a Lhasa Apso about how overloaded they are. We get it. When you hire the right marketing leadership *this tug of war goes away.*

That bears repeating: When you hire the right marketing leadership this tug of war goes away.

I believe I've gotten more done in the last few months working tightly with our IT staff than anyone would have believed possible because I know where MY threads fit into the grand scheme, when they can be inserted, and how long they will take to process and be replaced with new ones, or the paused older ones.

Ironically, yesterday (alas time does not stop when one writes a book) I was given leadership over the product road map, user interface development, etc., of our solution. Some might balk at this because it blends too much product management and technology into marketing, but I am very pleased. Reasons given included:

- You're on the front-line of customer development, Doug and you have the most direct line to the customer.
- You can interface with our IT guy very well. (Not a department, mind you, but a very dedicated single person.) You speak his language and know what's what. (Remember, I know enough to set expectations and manage results but in no way can I write a SQL query.)

MultiThreading Lessons from The Godfather:

During his first trip to Sicily, Michael Corleone finds a girl who catches his eye. He asks his escort team about the girl, her family, and the culture. By combining what he learns with what he knows, he finds that by winning over the girl's father - he can win the heart of the girl.

He didn't have to learn everything, but just enough to get the job done. By understanding the culture and human nature, he was able to "interface" with the family to get to his desired outcome.

CHAPTER 11

Sorry Dave, I'm Afraid I Gantt Do That

No Really - I Don't Hang Out In the Computer Section of Barnes and Noble

Hopefully you aren't tired of my pun-laden chapter titles yet. You try and work "Gantt" into one and see how you do.

I bought the book. I had the software. I took the online training. I was fascinated with Microsoft Project at some point in the early 2000s. We really didn't have anything better and I was in charge of a large project with a bunch of moving parts. My goal was to finally document and map what had been a "cluster" numerous times before. I finally got to a point where I understood how to make tasks dependent on one another, i.e., when one person fails to do X, then Y gets pushed out accordingly. Then I asked, "Well, what about giving everyone in the company and the external resources and partners on the project access to this Gantt chart and the guts of the project?"

I quickly realized after a little research and a few questions that this was basically impossible for a "normal" small business. We needed "MSFT Project Server" and more seat licenses and development resources and...the idea was dead.

Right about that time the term "online collaboration software" was starting to really emerge and after having the exact same experience described above with MSFT Share-Point, mine eyes had seen the glory of the coming of the cloud.

Your Big Ol' Task List in the Sky

Anyone in a larger enterprise surely knows the angst associated with Gantt charts. If you've been forced to use MSFT Project and set up all those dependencies only to determine that the chart was irrelevant and more of a time investment than the project execution itself, I feel for you.

Today's business landscape provides the MultiThreader with a vast array of collaboration and project management tools that are essential to her operations. Repeat that: *essential.* We all know how difficult it is to manage life when it's just your own few tasks and family matters. But when you begin to "see The Matrix" you will quickly be overwhelmed without a system for organization, account-ability, transparency, and collaboration. I use Central-

Desktop.com (1). I will let you know that I've tried all of the popular systems out there and I can say that Central Desktop has the right blend of power and ease of use for me. You decide on your own.

Central Desktop (CD from now on, ok?) allows me to manage my own projects with their associated milestones and tasks.   Additionally, it allows me to task out to the eight resources I'm presently using and gives me a clear picture of how things are coming together. Do you see how leveraging four affordable external resources on a project running in parallel threads can supercharge your results yet?

Currently, I have about 18 top-level projects and pieces being executed in parallel and 12 external people have different pieces of them.

Plus, I have people inside the organization working on some pieces too.   Remember my product road map responsibility? We have our IT projects on CD too. Now when we have internal wishes for the software, or we get customer requests, we put them on a specific task list for our IT guy with no due date. He reviews them for quick fixes, critical issues, etc. Then at our product development meetings, we sort through items, prioritize, then post to appropriate task lists, and assign due dates, etc.

Now, our sales team has joined in the CD fun and has deployed its own sales intranet portal so all of its resources, FAQs, videos, training materials, etc., are in one place with version tracking. Done. Total investment for a non-tech guy learning the system? *About four hours and he's 100% independent of IT and me.* Do you see how powerful that is? Taste that for a minute.

We feel completely in control of our destiny because of this organizational tool. E-mail has become less burdensome as conversations about projects move into "discussions" inside the workspace, archived indefinitely for our own learning and review. Files associated with projects or deliverables are uploaded and version tracked in the project or department workspace so there's always a safe source and no need to track down a presentation on a shared drive or inbox or local drive or _____ .

I want you to be comfortable living inside a system without feeling like it's a prison. I still use Google Docs and Calendar for quite a bit, even though CD has online documents and spreadsheets built in. Google Docs is just too fast and simple for me to sit down and knock out a script for a video, or share a spreadsheet with one of my external team members. If the document proves later to be a bigger part of the project and something that deserves to be a part of the record, I upload it and dump it from Google. You need to be flexible and if you live in MSFT Word and Outlook, so be it! (I'd suggest you use the Outlook plugin for CD too.) Just organize your projects, responsibilities, and key discussions about those projects in CD because your brain won't be able to hold all the stuff you encounter in MultiThreading. Try as you might, eventually you will feel overloaded. Don't do it. Collaborate remotely and feel your power multiply.

Central Desktop also has built-in online meeting capability so you can really amp up your remote team and remote resource collaboration.

MultiThread Lessons from The Matrix

Spoon boy: Do not try and bend the spoon. That's impossible. Instead, only try to realize the truth.
Neo: What truth?
Spoon boy: There is no spoon.
Neo: There is no spoon?
Spoon boy: Then you'll see that it is not the spoon that bends, it is only yourself.

In the above scene, Neo is still struggling with the system being a prison. This scene seems pivotal in helping him realize that the system, which earlier in the movie Morpheus describes as 'the enemy,' can be manipulated by knowing what's possible.

Again, I encourage you to become comfortable living in a system, and know that the system isn't your prison. Rather than trying to change the system, change your beliefs about the system.

CHAPTER 12

Blogging Wonkers and the Content Factory

"You're Really Willing to Admit That?"

Some dude actually said that to me as I walked passed him in the Costco parking lot. (He saw my Iowa license plate that says, "I BLOG".)

I Promise I Don't Live in My Parents' Basement

By now, blogging is not something new and exciting. Blogging isn't widely perceived as something negative and sleazy anymore, either.

Blogging is simply requisite for businesses as a way to quickly and efficiently engage in two-way and key word rich exploratory dialog with customers and target markets.

This book is not a "how and why to use modern marketing technologies". There are at least 50 of those by now and some are very good. I've recommend a list if you're looking to get the complete picture of why, how, and what tools to use on my References page. But at least semi-frequent blogging by your marketer is key to understanding his style or approach.

Experience in writing blogs will not only be a great source of research for learning about your candidate or how your current employees are communicating "out there", but it will also be a clear indicator that the marketer has:

- A basic understanding of grammar, language, and elementary HTML code
- The ability to engage audiences and have two-way conversations
- An understanding of basic blogging technology
- An idea of how sharing quality information and ideas can lead to inbound marketing
- An understanding of how to properly link out to others to increase exposure (which invariably leads to increased exposure for you and your company)
- A basic understanding of analytics and how to track visitors and inbound links

I started blogging in 2005 after my good friend Isaac Garcia, CEO of Central Desktop (www.central-desktop.com) explained what blogging was to me (albeit very roughly and before my move to Iowa). I didn't know then what power, reach, and pleasure blogging would bring me. Blogging opened the door for me to publicly share my passions, my thoughts on business, and my personal moments of clarity (now shared at douglasE-mitchell.com).

Here's my first blog post ever, from 2005:

kids. wow. one can't really explain the joys and the miracle bestowed upon steph and I with our two little ones. it's truly overwhelming. (prelude)

begin flight from orange county airport (SNA) to Chicago (ORD) en route to Des Moines (DSM). 15 minutes into the flight, gavin pulls out the binky from his mouth, let's out a short cry, and proceeds to vomit profusely all over himself, the seat, the dvd player, etc. shocked, i scramble to "catch it" w/my hands and i begin to wipe. steph yells "blanket" or something and about 5 minutes later, it's done.

oh god help me. my son has made a puke fest out of row 33. @#$@$%#^^%$^t. what to do. next thing i know, baby 2 is in someone else's hands, and steph's in the bathroom taking off boys clothes, and washing them. he's traumatized and the only worse trauma is mine.

they come back, he's in a diaper only. and for the next 8 hours, he remains that way, in airports, cars, etc. poor guy. had a fever for 2 days after so not sure if it was motion sickness or a flu bug.

i will never be the same…but this episode simply reinforces that i would literally die to take away the pain of my kids. i felt so bad for him, and would have done ANYTHING to bring back my spirited little trouble maker who worships the ground i walk on and waits anxiously every day for me to return from work to play on the pillows or outside. someday, i'll be telling him about how i held him in one arm, cleaned his poop, and how he was so cute when he said his own name for the first time.

what a gift.

I've read this post back to myself at least hundred times. Granted the post has nothing to do with business, but it probably gives some insight into "me". Blogging provides a simple vehicle to engage the world. Whether one intends it or not, blogging puts much of the essence of who we are on display (for Google and forever).

My blog which obviously began as a personal journal became a repository for business thinking and my commentary on moving my family from a lifetime in California to Des Moines, Iowa. My blog has:

- Led to my thoughts and stories being linked to by authors as a "source"
- Led to being invited to write articles for other blogs and nationally recognized publications
- Provided me with a vehicle to distribute my first e-book which has had thousands of downloads, subscribes, and views
- Led to people contacting me out of the blue and telling me that my writing about Iowa gave them the comfort and understanding of life here in the Midwest convincing them to pull the trigger and move here too
- Become a living history of my perspectives, learning, and thought streams for potential clients and even employers to review.
- Led to speaking engagements that have led to more speaking engagements that have led to business deals.

Business Case for a Business Blog

Blogging in business will absolutely give you an edge over your competitors who are not blogging. For excellent supporting data on this statement, please spend some time on HubSpot.com. Download their myriad white papers and research studies.

I can boil down their data into a very straightforward nugget. Businesses that blog frequently receive far more inbound leads at a fraction of the cost of those that don't.

Inbounds Leads are defined as leads (people) who actively took the step to raise their hands and say "Please market to me!" based on your company's providing them with something of value (white paper, case study, research paper, special report, etc.). They may have filled out a form online, submitted their information via a contact form, or picked up the phone to call you.

You must be diligent about measuring all types of inbound traffic to assess your return on investment, effectiveness of programs, etc. Using specialized 800 numbers, Google Voice numbers, landing pages with unique URL's, and special codes can be excellent measuring tools.

When I'm speaking, someone in nearly every group still asks, "How frequently should I blog?" There's no correct answer, but one thing is for certain: if you put out relevant, keyword-rich content more frequently, and have a means to convert that traffic into leads, you will by design or by accident have more chances for success than those who don't blog frequently. In other words, you're increasing your odds.

Some of what I felt were the most brilliant posts I've ever written have heard crickets across the Internet. However, my posts that solve certain problems for people or teach them the "Ultimate Prime Rib Recipe" have succeeded wildly.

How the MultiThreader Produces Whilst Others Formulate Excuses

As a MultiThreader I have a strategy and content production plan that keeps me on track. Here are some the pieces of my plan that we're executing now:

- **Set up a Diigo Group that everyone in the company, and our external resources belong to and contribute to.** If you've not discovered Diigo get started now. Diigo (Diigo.com) is a social bookmarking tool. With the Groups function, any article we found through Google Alerts, through surfing, through RSS feed reading, etc., can be bookmarked, tagged, and shared with our internal private group. This is an incredibly efficient way to share information because it's a "safe source". E-mailing articles to distribution lists amounts to corporate spam. There's an amazing "auto-blog" function in Diigo too that once set up, means that simply bookmarking an article, adding some commentary, and tagging it right from your browser, creates a blog post automatically on your site. Very slick.
- **Assign each person in the firm 2 blog posts per month of 250-400 words.** Team members are encouraged to discuss issues they've encountered during the sales process, trends they see, etc. Although there's

some initial shell shock and concern over "having something to write about," this quickly fades and people look forward to having their voices heard. After all, a company is simply the sum of many human beings so why not show the world the humanness of your firm? For us, this means 32 posts per month OR 11,200 words that are indexed by Google. If you've started at the beginning of this book, you're right at about 12,000 words and trust me, it's taken me far more than 30 days to get this far. We have an internally published "content calendar" so there's totally visibility for due dates and expectations.

- **Leverage local content ninja freelancers to build additional content.** Currently, I work with 3 external resources to affordably build content for our blog. Some would cry "Inauthenticity!" and to them I say "Get over it." I don't have articles ghost written that espouse our brand, culture, and intimate feelings about topics. Rather, I leverage external resources to stay on top of national trending topics, niche content pieces that spark an opportunity to build value for our audience, and micro-niche content that targets three and four word/keyword combinations that I've researched in Google keywords and want to dominate in search results.

- **Leverage a PR freelancer (non-agency) to create longer format white papers or case studies each quarter.** Sure we produce interviews, testimonials, case studies and white papers on our own, but we have quarterly pieces being worked on externally so they get the attention and focus they deserve. In our fast paced marketing life, we often must sacrifice detail or depth to execute. This method provides a safety net so we don't "under attend" to something critical. We also leverage our PR freelancer to write more traditional releases and push them into editorial circles. We're lucky in Des Moines to have a rockstar who has experience in our industry. She developed a robust media list as a first step so we know exactly what our reach is. Our first release, by the way, generated three interviews for trade magazines and two offers to write articles. Most every trade magazine fills its pages with material provided by companies like ours, so take that with a grain of salt, but every chance to get your name into the marketplace is a victory.

The net result of this content plan, in addition to all of the other ad hoc content creation that happens, is that we're casting a very wide net. Leverage content ninjas and external affordable resources to cast a wider net and all the while, you'll be building up more people in the market who know your company, your target markets, your culture, etc.

The master of content-driving leads is absolutely Hub-Spot.com. The company generates ridiculous amounts of inbound leads by producing superior white papers on relevant topics and holding very useful webinars. But get this; they have 170 people on staff. Granted they built their business smartly with content all the way, but you and I both know that with a large staff and gobs of cash we could create awesomeness too right? So just remember, create content on as massive a scale as you can and don't believe that no one in your industry could write or understand you well enough to add value. There are plenty of good writers out there who will add value and start giving you fresh new ideas from their outsider's perspective.

"Blogging" Lessons from Steven King

Stephen King was rejected on a scale that would make even the kindest pet rise up from the grave and unleash kitty hell on one's psyche. King's *Carrie* was rejected 30 times and he eventually threw the manuscript in the trash. King was writing and writing and writing, and no one seemed to care. It can feel a bit like this when you start blogging in your business. It can feel like the universe is rejecting your manuscript.

As King did before you…keep after it. Find your voice. Make a Stand.

CHAPTER 13

Content Management Syndrome

"But Isn't That Just Blogging Software?"

I can't tell you how many times I've answered that question about WordPress, the free and thoroughly amazing content management system that makes life easy for millions of agile businesses. "What's the difference between a blog and a website?" usually follows.

My answer: Nothing. Of course, blogs are more typically a flow of content in categories over time while pages are more static but Google doesn't view sites built on WordPress as "blogs". In fact, Google loves the awesome "SEO Structure" native to WordPress sites. The use of plugins has further extended the richness of WordPress to turn most wishes, wants, and desires for functionality into a few clicks instead of a few thousand dollars in custom code.

WordPress Is Like The Force for Non-Jedis

WordPress started out as blogging"software, meaning that it made updating and adding pages and blog posts simple and straightforward for the masses. Its editor makes updating our web site and pages as easy as sending an e-mail. The explosion of free or affordable templates has made WordPress into a powerful website solution for companies small and large.

Our Custom CMS Is Way Better than WordPress (and $500 more per month)

I still see companies deploying custom "Content Management Systems" or CMSs. These CMSs were largely built by ad agencies and custom software houses as alternatives to those open-source pooh-pooh solutions that the unwashed use. Blurg. In my experience, WordPress has easily surpassed the functionality of any custom CMS I've ever encountered. Even if it lacked in comparison (which it doesn't), the cost and productivity savings using Word-Press make its use a slam dunk.

The cost of developing these CMSs must be recovered at the client's expense. The cost of maintaining outdated and bloated code is becoming absolutely unwieldy.

Do you know what's happening everywhere? Agencies are figuring out that clients don't want $40,000 websites with custom CMSs. They want $5000 sites or *one* day design-to-live micro sites built via multithreading on WordPress.

WordPress Ninjas are always busy deploying small customizations and integrations between solutions today.

While free and easy, there's always something you'll want that goes beyond standard. But you'll find that once you invest the time to understand just enough about how WordPress works, your problem will be remaining satisfied with your current set up and not overloading your system with myriad plugins to add this or that cool functionality to your site.

The MultiThreader's Perspective: Quality Content Trumps Tech

The MultiThread Marketer knows that having an affordable and effective content management system in place is critical to the rapid dissemination of relevant information to its target audiences. One of my first initiatives in my new position was to deploy a WordPress site to complement our main site. I also took over all of the static pages on the main site by pointing those addresses to new and customizable WordPress pages that I managed. In other words, I took the brochure pages away from the land of IT and gave the power to the people.

Now changes are made on the fly. Pages are developed in minutes. We can push out a version one in the morning and be on version five by lunch. Adding someone to a team page is a few minutes of non-coding ease. Previously, this would have been a multi-day, get to it when I can, plunge into IT hell. No more. You're liberated.

Plugins Turn Ordinary Humans into Coding Rockstars

Plugins are little bits of code that make life easy. You've heard "There's an app for that". Well, there's probably a plugin for that too. As developers uncover functionality that doesn't exist in WordPress that would save them time and energy, they build plugins and distribute them freely and sometimes ask for a donation. Some of the most useful and amazing plugins can be found in this Podcast: Managing the Edge Episode 21: Essential WordPress Plugins - http://managingtheedge.com/episodes/essential-wordpress-plugins-episode-21. There's very little that cannot be accomplished by mere mortals with plugins. FYI: You cannot use all of these great plugins on the WordPress.com platform. That's hosted by them. You can go fiddling with the machine, but you can only use what's approved by them. You must use the hosted Word-Press.org platform (for perhaps $5/mo. on Godaddy or some other cheap host) to experience the mega-awesomeness.

MultiThread Marketing Lessons from Jim Spanfeller

The future of value creation in the digital world will almost certainly be more about content than technology. The platform is starting to mature. We have arrived at critical mass and open-source platforms, and the general transparency of code makes replication all too easy.

In reality, the world is almost exactly the opposite of how many see it today: It is the technology that is commonplace and the quality content that is unique. What a shame that many of today's content companies will not be around to profit from the sure-to-come better understanding of the digital ecosystem. Why? Because they, themselves, are among the core promulgators of these misperceptions.

—Jim Spanfeller

CHAPTER 14

Social Media Madness

Don't Mistake a Lack of Software Fees for Free

I've spoken to a couple thousand people on the topics covered in this book. None is more commonly misunderstood, asked about, and maligned as Social Media. Invariably, I still get asked the question with the follow up, "How to you find the time to do all that social stuff? I'm too busy doing business to waste time on that!"

I make the point very plainly that like many mediums that met resistance in their early days (telegraph, fax, e-mail), social tools are merely an extension of what my audience already does. Have you ever referred someone to an open job? Have you ever had lunch with someone and turned her onto a connection that opened up a huge new business opportunity? You were social networking - in person.

Now we've simply moved some of that interaction online and exploded our potential reach.

If You Call Me a Social Media Expert I'm Going To Punch You

Of all the chapters in this book, I put this one off the longest. That's probably due to the over use and abuse of the term "social media" and what it means to most small businesses.

One of the most overused and misunderstood terms in business these days is "social media." Because of the quick entry into massive use, tools like Facebook, Twitter, and YouTube, many business - those using social media or not - aren't tapping into the business potential.

Social Media, at this writing, either draws a crowd of misunderstood outsiders and naysayers, or has captured the confused and trapped into a time-suck.

One of the rally cries of Social Media cheerleaders is "The relationships matter most." I'll agree that relationships matter. And I'll even go a step further and agree that using Social Media tools should be the practice or intent to improve and create relationships - but with an eye towards creating more business, or more customers. Never lose site of this fact: Twitter followers, Facebook Likes, or comments on a blog will not keep your electricity turned on. Social media tools are merely a tactic that typically fit into your overall strategic marketing plan.

Social tools prepare customers and prospects to buy whatever it is you're selling, be it a product or a service. They can also be a fantastic tool to create relationships with your future brain bench, and influencers of your future customers, as well as potential vendors, investors, or collaborative partners. This is known as Proactive Bench Building.

For the general public (and many a "social media consultant"), the feeling is that social media is to be a day-after-day chit-chat, what's-up-with-that, Seinfeld-like conversations about nothing (related to business).

The MultiThread Marketer, however, has a purpose for every message sent. And every message should be measurable in such a way to tell if the practice is building business or customer-base.

Strategy Before Tactics

Put simply, strategy is the goal, tactics are the steps towards reaching that goal. Making (or attempting to make) a viral video is fun, and may actually go viral. But does it help reach the the goal?

If you want X number of new customers, figure out how many more phone calls, walk-ins, or appointments you need to fulfill that. Once you have that number, that becomes part of your strategy as well as a measurement to see if your social media plan is working.

For example: A restaurant is doing a great business during dinner. Every night, it has up to 30-minute wait times between 5:30p.m. and 8:00 p.m. Why should they use social media? Well, they could either upsell during the busy time (is there a packaged product they can add, or a menu item that will increase the per head average?), or they can offer unique specials for early-birds or large parties during non-peak hours (think volume).

In the above example, the strategy is in place, now the tactic is to talk towards those goals. You'll also be able to measure the success of your social media use, because you practiced strategy before tactics.

If you're uncomfortable writing - get comfortable. If you can talk, you can write. Still, if you're still hunting and pecking at the keyboard (see Chapter 1!) or whatever lame excuse you might have for not writing, outsource this piece (Multithreading, remember?).

Another tactic in social media, is to connect with influencers. Influencers can help spread the word on your business. This is a form of Multithreading, no?

And just how do you influence the influencers? Know what they're looking for (resource or information, not gifts), and give it to them. By adding value to their lives, they will often reciprocate with giving you and your business due props.

The B2B or B2C Question

In my experience, leveraging social networks to gain sales in a reasonable time frame is much easier in Business to Consumer markets (B2C). Social networks are far more likely to act on tweets offering discounts, specials, or give-aways than they are to hearing consultants distribute sage advice in 140 characters.

My over-simplification is that social media in retail can really pack a punch and the same tools applied to consultants and manufacturers of boring but necessary things have a far different "meanginfulness curve". When I get a question about this during a talk, I can best answer it in the following way:

Well, for a small business like mine, the answer should be obvious - I'm standing here today talking with you about becoming findable online and leveraging these tools to win, right? If you go to Google and search for me you'll find my Twitter.com/mitchgroup profile at the top of the results, right? You haven't thrown tomatoes at me so you're believing what I'm saying, right? Well, how do you think I came to stand before you in this very moment?

I put myself out there, engaged, wrote, created video, built brand awareness, and invested in an online representation of who I am which people seek out, find, and use as a reference point. Then they choose to either have me come tell yet more people about it, or they hire me. In B2B social media is a credibility and brand awareness tool that requires investment but like many things, the payoff isn't typically instantaneous. It's a long term deal and it pays serious dividends.

In 2007 when I went out on my own, I knew very little about these tools and how to leverage them to win. But I invested and kept after it, and now I'm getting paid to stand here and tell you how to do it. Is it worth the return on investment? You decide.

Analytics on Social Networks

I've had many a heated discussion on "The ROI of Social Media" (or of PR for that matter). How can I prove that my 1000th tweet that has taken 18 hours of my life over a year was the one that made you buy from me? I can't. Similarly I can't prove that 18 articles refreshing my name in peoples' minds didn't prime them for the buying decision. In-depth research and confirmation of how people found you and exactly why they bought can be hit and miss. I prefer to look at things this way. In B2C we can measure that a tweet offering a $10 bunch of roses today between 1:00 p.m and 2:00 p.m delivered 25 orders. We can measure the demographics of our fan base on Facebook as being 80% 18-34 year olds and say, "That's who we're targeting and that sector is growing." In the B2B world, we have to factor in brand reputation, brand awareness, the reach we gain with our materials we distribute, e.g., a case study that gets wide distribution on Twitter that leads to an interview with a thought leader that leads to a consulting engagement in your town. Different path. Same result.

Don't OD

I'd like to close this chapter by warning you to avoid the pitfalls of most social media types early on: overload. Believing you must be on every platform out there and always on is a joke and it will make you a shallow web thinker and likely cause burnout. Pick a platform or two that make sense for your business, and invest. Niche audiences and groups are where people are headed these days because the noise out there has become too great to manage. So many great tools blow up into something big and worthless. Pay attention to hyper-targeted niche social networks and groups that filter noise on purpose and be selective.

Differing Opinions of Social Networks

"I despise Facebook. This enormously successful American business describes itself as 'a social utility that connects you with the people around you'. But hang on. Why on God's earth would I need a computer to connect with the people around me? Why should my relationships be mediated through the imagination of a bunch of super-geeks in California? What was wrong with the pub? - Tom Hodgkinson

"Social media changes the relationship between companies and customers from master and servant, to peer to peer." - Jay Baer

CHAPTER 15

Paying Less For Organic

How Many Salespersons' Quotas Did I Ruin Today?

Honestly, as someone who's been in both inside and outside sales roles engaging Main St. and C-Suite clients, I don't enjoy making sales people squirm. (OK - maybe a little.) Recently, I was asked to sit in on a meeting between our client and a local TV station's online sales team to give my input. After pleasantries, the rep said, "So we can get you at least 500,000 impressions over the next month and that will really help with your branding!" Now imagine the sound of screeeeeeeeching brakes. Errrrrrrrrrrrhhhh!

Me to the rep: "Branding huh? My client's brand is over 80 years old. The company is in the top 10 in its industry and is known statewide as a leader in its market. We're not interested in branding and impressions. We're interested in click through and conversion into orders!"

Again - meeting adjourned.

2009 Google Ad Word Revenue was $23 billion. (1)

The MultiThread Marketer needs to have at least a macro understanding of the world of organic results versus paid results *(sometimes called PPC - pay per click or SEM - search engine marketing)* in the search engine universe.

Organic results are effectively what show up in the middle section of the screen when you type in a search term. Paid results are the "sponsored ads" that are on the top and sides of the page. The difference is pretty obvious.

You earn your way to the top of organic results and you can pay your way to the top of PPC results. I'm over simplifying here, but you get the idea. (For more on PPC advertising and how to do it right, visit http://managingtheedge.com/episodes/the-best-payperclick-and-adwords-podcast-youll-ever-listen-to-episode-24.) Your mind will be blown when you're done but this show lays out everything you need to know.

Too many times, a CEO or other company leader says, "Hey, go buy some ads on Google and let's generate some traffic so we can make more money." The marketing department sets up some ad campaigns with a daily budget and they're off and running. The result? A complete waste of time and money.

Clicks for the sake of clicks are meaningless and the MultiThread Marketer knows this.

In my experience about 10% of businesses know what their specific outcomes are when they embark on a PPC campaign. "I want to sell 25 more widgets per week." "I want to generate 35 new sign-ups to my e-mail campaign every day." These are **specific** examples as opposed to the general "We need more traffic!"

We can narrow down even further by the businesses who understand that a searcher is typically looking for something specific so when he lands on your page after clicking your ad you'd better deliver.

The MultiThreader knows that depositing a search on the home page in hopes that someone explores our site, absorbs the brilliant light that emanates from our words on a page, and either buys or signs up is absurd.

PPC is a little art blended with quite a bit of data and science, so as an accomplished marketing leader don't do this by yourself. You understand that toying around with a tiny PPC campaign *might* be doable on your own. In fact, I'm testing a couple very small campaigns now on with a couple under-served and low competition keyword combinations with a total exposure of $6/day. These are my "pet projects" as a supplement to our full blown campaigns, run by the experts.

A full-fledged campaign (one with many sub-campaigns under it) is fairly complicated to set up and is too much for you to manage day-to-day. You will want to take a look at results every week or two but you should start any campaign with a monthly budget of over $500 with a reputable PPC/SEM firm or individual in your area.

Using an outside resource will keep you from falling into these PPC traps:

- Using keywords that are too broad which eat up your budget with worthless clicks.
- Not setting up your campaigns properly for maximum quality score. (Google ranks every ad with a quality score by a set of criteria.)
- Not using exact matches to exploit "long tail" keywords combinations of 3 or 4 words that really capture the searcher's intent and land them on an extremely relevant result.
- Trying to spread a small budget too thin so you end up with poorer results than if you honed in on high value/high conversion ads.

Typically, PPC/SEM firms charge a set-up fee up front the first time you engage them and often charge about 10% for ongoing maintenance of your campaigns. I've encountered firms that simply put certain key terms on auto-pilot for small businesses and do very little to manage the account. The firm sits back and collects its management fee and says, "Well if you leave us, we'll find another (fill in type of company) in your town and make them show up on top so you better not leave."

Companies like this are merely holding you hostage because frankly, anyone can outbid the competition on PPC ads. Yes, better quality score ads CAN show up ABOVE those willing to pay more for the click but again, this isn't rocket science. PPC is a known commodity. Delivery of PPC services is NOT a commodity however, and companies who provide the service must be researched.

Ask for referrals in your network and of course leverage your social networks but be warned - if you ask for referrals on Twitter you will immediately be deluged with new followers pimping their wares on you. Be cautious and perhaps try using good old-fashioned in-person networking and referrals first.

**Be very careful to get references from people you trust on SEM firms. There are few consultants and firms out there who *really* get SEM at the scientific level needed to make an impact on your business. There are many consultants and firm out there who know just enough to sound very effective because you haven't done your homework.

Going organic is NOT free

Organic results, i.e., those in the middle of the page, are generated by a completely different set of results. Google has a magical algorithm that decides what's relevant to the universe when someone types in a key term. Relevance is determined by many elements and Google can change the way this works at any time. Moving target? You betcha! However we know a few things that are pretty stable about getting organic results. We'll avoid the nitty gritty here about how to structure content specifically for results but there is plenty of that contained within the podcasts over at www.managingtheedge.com. Here are some basics.

The keyword or phrase the searcher typed in is contained in your online content (page, blog post, etc).

If someone types in "How do I floss a cat" and you have an awesome article you've written called, "How do I floss a cat", your chances just went up of being found and high up on that first page of results. Notice I didn't pick a keyword phrase like "Social Media". If you never want to be found on the first page of results, only talk about topics that the entire universe is talking about. Odds are you won't overcome the masses and the noise. The keyword phrase I chose is very specific and it's structured in the way someone might search for it. Keyword combinations that are three or more words and structured in the same way a human might search, increase your chances of getting organic results dramatically.

Getting people to link to your piece of content, "inbound links", is highly desirable and factors heavily in Google's "relevance" determination.

Why wouldn't inbound links boost relevance? If 1000 people link to your article from their blogs, websites, etc., and say, "Wow you HAVE to read this post on cat flossing!" it's a good bet you've got something. Google rewards you by pushing you up the page because after all, you have what people are looking for.

There are ways in which marketers can attempt to game the inbound linking system but I've never used them, and Google continues to ferret them out and punish people for using them. The general rule that relevant content that people like and link to rises to the top is not likely to change.

Niche content has a far greater organic chance than mass market content

My keyword combination was pretty wacky but I chose it to illustrate the point that your business needs to find a way to become relevant to many micro niches instead of an entire sector. You'll have far greater success developing content for passionate online searchers looking for quality content than trying to become front page results talking about the economy. Perhaps talking about "How the economy has affected the pet food market" or "How a down economy boosts cheap dog food brands" could have a much greater chance to produce results for you.

MultiThread Marketing Lesson from Isaac Rabinovich

"But by trying to make their ads less obnoxious, Google removed all the visual cues that these cognitive filters rely on. Which is why market research indicates that most people don't perceive Google ads as ads, even though they're clearly labeled as such! In other words, Google found a way to get past people's anti-ad wetware." —Isaac Rabinovich

CHAPTER 16

Warning GRAPHIC Language Ahead

Run. Run Away. Fast.

Me: I'm pretty sure I know what I want. I need a logo and a header image for a new blog site. I need a background image and I'd like earth tones with some navy blue and burgundy accent colors. The logo image should be square in dimension and needs to indicate academics, learning, or knowledge. How much will that cost me and when can you have it ready?"

Him: Well, let's schedule a consultation session where we can brainstorm ideas first. I really can't quote you a price when we don't even really know what you're looking for.

Me: I just told you explicitly what I wanted and even gave you color palette ideas. I'm not asking you to search your feelings about this project. I've asked you what it will take to execute it. This meeting is over.

This is a true story. Remember, most often, artists are poor business people.

Would you hire a painter as a project manager?

Too many companies wrongly perceive their marketers as "the people who develop the logos, brochures, and coffee mugs". It's not often stated overtly, but a high percentage of agile companies still hire Marketing Directors and VPs of Marketing because they "know how to use graphics software." Seriously!

It's not hard to understand why. Companies believe in their DNA that the marketing department is in charge of their "brand". Brand is still viewed as "looks" instead of the complete customer experience. Companies hire based on their DNA and voila we have a Marketing Director who spends her time making buttons and banners.

The MultiThread Marketer knows what looks good but understands that looks are a means to a conversion. Good design and a good UI (user-interface) guide the visitor to become a user, consumer, buyer, lead, prospect, or consumer of information with as little resistance as possible.

Graphics are in essence the WD-40 that smoothes the path from "search" or "browse" to "buy".

When seeking out your strategic hire, make sure she has freelance graphic contacts in your area. Unfortunately for graphic artists, the market is pretty much always flooded with talent. The kind of talent, that understands *business*, however, is much tougher to find.

I pay anywhere between $15 and $85 per hour here in Des Moines for graphic design services. When I have the time and need simple things that are merely executed according to my instruction with multiple revisions, I can afford to stay at the low end.

When I'm doing mission critical work requiring a broad understanding of my business, my goals, my desired outcomes, and I need output that hits 99% of what I want on the first try I'm at the high end of the pricing spectrum.

There are graphic designers who fancy themselves worth $150/hour because of their depth, their ability to capture "who you are" with a swoosh and a color choice. In my experience, this is wasteful.

If you find someone who is absolutely dead on in 2 hours at $150 each, of course it's better than floundering for 6 hours at $85. My experience shows me that at every price point there are excellent people who execute with precision. MultiThreaders know who's who and what's what.

If you are striving to hire a MultiThreader, he must have these resources on tap. Otherwise your candidate will likely default to using an agency that can just do everything and wrap it up for them in a nice package.

I currently leverage three different graphics ninjas in town at three different price points. The MultiThreader needs a bench of talent on tap in case one is overbooked. She proactively builds that bench at every opportunity. A great way to keep building the bench is to give sample projects to many people. Test how well they grasp your business and what you're trying to accomplish. It's absolutely worth a couple hundred dollars and some of your time to uncover a rockstar resource. This is purely R&D so don't try to justify your valuable time.

Maximizing value

Your marketing leader should know a few basics and some hackery regarding graphics formats. For example: If your marketing leader has a local graphics freelancer build a banner for a landing page, he should know how to ask for a "reusable" banner with a place for replaceable text boxes that he can edit without hitting up the freelancer each time a change is needed.

He should know the different file formats and what software opens and manipulates them. Simply knowing that the open source graphic editing alternative GIMP is your ticket to editing .EPS graphic files. Knowing that .EPS files are extremely common in the graphics world as high resolution scalable files that most printing shops use, will save gobs of cash.

I recently began a targeted landing page campaign and needed some 860x200 pixel wide banners and I specifically asked for the key elements of logo and tagline but wanted to be sure that there was space for replaceable headlines. The file came to me as an .EPS file and I use GIMP to edit it.

I've built nine variants of this banner so far. Total time investment for me: 22 minutes. Total outsource payment $22.50. If you're saying, "Why didn't you just have the external resource build all 9 variants at that price?" You're thinking like a MultiThreader! However, in my case, I didn't know what the variants would be. I didn't know I needed nine. I just just knew I needed the right dimensions and a base from which to work.

Also have your brochures or other materials built with the same principles in mind. One thing is for certain, you'll want to change whatever piece of printed literature you produce 10 minutes after you receive it! So have your materials that you'll be printing yourself produced in a way that lets you modify the text inside without requiring the graphic designer.

You may find yourself paying a high hourly rate to absolutely nail a brochure for your sales team, and then find that you're paying almost nothing to have some changes made to it by another graphic artist. You can decide as you move along whether doing it yourself is worth it. Perhaps changing the a to an A is easy enough. Just watch yourself! If you are graphically inclined as a Multi-Threader, it's easy to get sucked in and start redoing, redesigning, and squandering your precious time.

OK, Captain Production - Don't forget you're a leader!

Hiring a MultiThreader whose background is creative may require some deeper due diligence. Your goal is to discover what happens when push comes to shove. Invariably, marketing departments have one universal complaint. "We need more money!" In your agile company environment, there probably IS no more money. So if you've placed a creative or graphic artist in the leadership position and the workload is intense (and when isn't it?), you must be certain that she won't digress into a producer of artwork when she should be leading.

Of course there's a balance. In my work, I find myself spending time on things that I'm good at and enjoy doing from a creative standpoint, but I have the awareness of how long I can spend on these things, the realization that I'm doing them, and a specific outcome in mind. Plus, these things are usually centered on creating demand and giving our sales team another arrow in their quiver.

The MultiThreader has a keen awareness of what she's good at and how to build her passion and skill set into the marketing mix of the agile firm, without turning into a production house when leadership is key.

Lessons in Art by Donald Trump and Kermit the Frog

"Deals are my art form. Other people paint beautifully on canvas or write wonderful poetry. I like making deals, preferably big deals. That's how I get my kicks." —Donald Trump

"How important are the visual arts in our society? I feel strongly that the visual arts are of vast and incalculable importance. Of course I could be prejudiced. I am a visual art." — Kermit the Frog

Section Three: Making a Case (Study) for MultiThreading

Littered amongst the previous chapters are some of my own personal stories and some anecdotes taken from familiar fiction works. In this section, are some real-life business case studies that may help tie some threads together.

Proactive Bench Building

Investments Pay Dividends

I wanted to make some SEO and search enhancements to my company's website so I went to our developer and said, "Let's do A, B, and C". He promptly explained to me that life is not as simple in MSFT land compared to what I'm used to.

That meeting sparked my launch of a 2-month casual but focused search to find additional resources who know some less-used technologies by today's standards. I put out feelers and began having conversations with a guy on contract with a big firm in town with miles of older code. He then connected me with two people who are expert in the search methodology we use. The net result is that I now have an expert who is about 80% self-sufficient inside of my systems. He knows our processes and our technology base and he's able to walk up and dive in. Oh - and the search *wizard?   He figured out that we didn't have a search problem.   We have "communication problem" with some hardware and software at the hosting level.   Total investment so far in both resources - about $1200.   Future payoff - Immeasurable.*

Whaddaya Mean You Don't Have Ninjas Here?

As a MultiThreader, a portion of the job is to proactively refresh your resource pool and recruit new talent.   In essence, the marketing leader builds a bench of talent not unlike any other traditional business.

Early in my business, I invested too heavily in too few individuals and didn't understand how important bench building was until it was too late.   "I'm going to work for company ABC and can't work for you anymore."   That will REALLY ruin your day when you're in the heat of 8 projects of which 7 involve the quitting resource's skills.

Today I spend 10-15% of my time building my bench through networking, running test projects with free-lancers, asking for referrals, posting opportunities on Twitter and Facebook, etc.

I actively build a bench in the following areas:

- **Content Ninjas** - These folks can write articles, blog posts, and possibly white papers and special reports.   They can also typically manage social media engagement.   I hope to get a "90% right" ratio out of the gate on their content and take special note of how they're adapting to and learning about my markets, products, and clients.   In my market, Content Ninjas range between $10 and $35 per hour. You need to be test marketing for content constantly! Try someone out. Ask them for a sample piece on a topic after you deliver some bullet points or examples of other posts on topic. See what happens (and do this over and over again).

- **Online Ninjas** (Light Technical) - Online Ninjas can do just about everything using the basic graphic interface of most hosted or online applications. If you give them a word doc with 10 pages of blog posts to be scheduled over time, they can create blog posts in WordPress, apply the appropriate scheduling delay, and enhance the posts with images and tags. There's most likely very light HTML involved, but nothing that can't be gleaned from searching for "How to link an image with HTML." In my market, Online Ninjas range between $22 and $35 per hour.

- **Web Ninjas** (Heavy Technical) - These guys and gals can pretty much do it all from setting up an optimized WordPress site to customizing themes to managing database structure and hook ups. They know SEO and how to structure you for success. They probably don't want much to do with your IT infrastructure however so be aware. Fixing your printer and hooking up your WiFi are NOT things they typically want to get involved with so don't ask. In my market, Web Ninjas range between $50 and $95 per hour.

- **Internet Business Ninjas** - (The Holy Grail of Ninja-hood) These folks are web ninjas with business sense! They understand conversion marketing, content crea-tion for SEO, landing page construction and design, and e-mail marketing inside and out. They know how to do niche keyword targeting and have an eye for markets that can be exploited. If necessary, they'll dig into their tool kit for old school programmatic skills, but frankly most don't have the patience for re-writing Visual Basic code. They aren't "designers" but they have a good enough eye that they can make something pleasant and appealing without designing something in GIMP. If given a graphics file, they can chop it up into the necessary pieces and make it work with a template or theme. In my market Internet Business Ninjas range from $75-$125/hour.

I am lucky to have one such ninja in my stable. Let's call him "Cyclone" because if I tell you his name, invariably, he'll be too busy to do my stuff! Cyclone has plenty of ninja stars to throw at my projects and there's never a shortage of "You mean I can do that with just a plugin???" talk when we get together.

Remember that a couple Internet Business Ninja hours will often yield 10X results so think through what you really need before choosing your ninja.

I've also recently expanded my bench building into **Legacy Coders** and **Search Wizards**. Let's face it, your business may not be running on the latest and greatest web technology or code base. In fact, you're probably running in Microsoft land. That's fine but these folks are tougher to find because they don't typically hang out at coffee shops for the free WiFi. They're working large companies as full-time contractors. They're employees of Inatech Corp.

and punch a clock. You have to look harder for them but your search can really pay dividends.

The 6th Man

There are too many sports references that apply to this chapter where the victory points to the guy on the bench, the guy who practiced at 110% all year, only to be called up in the last 30 seconds of the game to make the winning score. You just need to know this. Ready?

ASK.

Always look for your 6th team member. Ask people around you for references to build your bench constantly. Ask. Ask. Ask. You shall receive.

CHAPTER 19

Repack Repurpose Reap Rewards

In my line of work, I often talk about the value of out-posts and repurposing content. This idea is not new nor is it mine.

The most basic example is to take 100 blog posts and create a book from them that costs $9.97. The book will have some extra material and enhancements that make it unique and different from the blog posts alone.

Next, record audio for the same book and sell that for $9.97. Include additional examples, case studies, or "audio only" special material.

These pieces now move from your blog only, into online bookstores and marketplaces.

You've taken publicly available content (your 100 posts) and repackaged them in a way that adds value for people and commands revenue. You've taken your passion and if you have something the market deems valuable to say, you've monetized the same darn stuff with little effort.

Managing the Edge as a Case Study in Content Outposts and Repurposing

In my Managing the Edge Podcast with my colleague Andy B., here are the inputs and the outputs. Notice the distribution leverage we gain through automation:

1. Do a live streaming show at a set time which provides a live audience, interaction, and word of mouth. (1 hour)
2. Show gets recorded automatically and is archived by Ustream.Tv our streaming host (free - no effort).
3. If short enough, the video is automatically pushed up to Youtube. (free-no effort)
4. I take the audio that was recorded during the show (while the video was streaming), push it up to Amazon s3. (10 minutes)
5. Andy creates a blog post that summarizes the show, links to key resources, and embeds the video broadcast and audio podcast into the post. (20 minutes)
6. That blog post goes out via RSS feed, e-mail blast, Twitter, Facebook, and iTunes automatically. (free-no effort)

I'm simplifying above but the reach that we gain in about an hour and half of work is amazing. Next steps for us might include episode transcripts that become an e-book, or wrapping every episode up into an enhanced audio course and selling it via internet marketing.

CHAPTER 20

Enlargement Dysfunction Nigeria

Who Buys This Stuff?

Thank goodness spam filters have evolved to catch most of the garbage that flows out of incognito servers from all over the globe. I guess someone must really be buying the erectile drugs, fake watches, member enlargement pumps, and so on that are pushed into my spam box daily. If it didn't work, they wouldn't do it, right?

I Swear I Sent It!

An unfortunate side effect of the spam industry is that you and I often get sucked into spam land even though we have a great message, high value, and honest intentions. To compensate for these high powered shields, we must develop a weapon of mass instruction! We must use technologies that maximize our chances of being seen. We must create messaging that gets read. We must respect the rules and ensure that we're gaining permission to e-mail a target/suspect/prospect.   And everything is working against us to achieve these goals.

Isn't E-mail Dead?

I've been hearing for years that with the advent of RSS feeds, micro blogging, direct messaging, and online collaboration tools, e-mail is dead. If you survey 1000 business professionals in small, medium, and large companies, I'd venture to say that over 80% still live in their e-mail application as the hub of all activity they execute.

As we approach the smaller business end of the curve, you will start to see a greater experimentation with other platforms, methods, and systems.   This is why even the coolest and most advanced collaboration tools have e-mail alerts and MSFT integrations. Google has embraced the "e-mail as a hub" concept and tied together its functionality to create an ecosystem that works with calendar, tasks, docs, etc., all within reach of the Gmail or Google apps browser window.

Double Opt-In Reigns

Double opt-in e-mail means that someone submits their information via a sign-up box and they receive an e-mail with a confirmation link. Once clicked and permission is confirmed, the submitter is added to the e-mail list and he begins receiving the material. This method is the safest and most powerful way to build large lists and to e-mail those subscribers without being tagged as a spammer.

E-mail marketing solutions like: MailChimp, Aweber, and Constant Contact all have relationships with Internet Service Providers who monitor their adherence to safe e-mail practices. These services force the issue of safe e-mail practices by building their systems to ensure they're used properly. If you buy a random list of 20,000 addresses and upload it into Aweber and blast away, I'd give you about 20 minutes before your account is suspended and gone.

All of these systems make it painfully simple to create an opt-in form, embed it into a site, and begin gathering subscribers (assuming of course that you are delivering some kind of value that people want).

It's Like Water-Boarding For E-mail

These systems also enable "drip marketing" campaigns to be automated and deployed in order to multiply your efforts. For example:

1. The subscriber submits her name and e-mail
2. She confirms the opt-in link
3. She automatically receives a welcome message with more value, an offer, some links to relevant videos, an e-book download link, etc.
4. 14 days later she automatically (and with no more effort on your part) receives an e-mail asking her how she liked the e-book along with more valuable content.
5. After 30 days, she enters a cycle of receiving your monthly newsletter that is automatically created from selected blog posts you write (yes, this is absolutely awesome and once set up is a no-brainer huge win).

Once you set the system and the initial e-mails up, you're automated. These drip campaigns will pay dividends because it puts the check-in factor on auto-pilot. Of course I'm not suggesting that you set it and forget it. In fact, I'd recommend a monthly quick check of your e-mails and content to ensure you've not broken any links or have irrelevant or oudated information in them. The more timeless the content the easier the long term management. The trade off is relevance, of course.

MultiThread E-mail Marketing Case Study

In createWOWmedia's work with Boesen the Florist (A Top 10 in the country by volume and revenue firm), we created an e-mail marketing campaign that is seeing excellent results. Full case study comes later in this book.

Identified Weakness and Goal: Boesen wanted to increase their annual wedding bookings by a substantial percentage and streamline the lead capture process. Their current .com website had an outdated wedding page that wasn't geared toward lead capture and Boesen was relying on their brand to carry them through and drive brides-to-be to call Tom Boesen to set up their consultation.

Tools and Technologies Used: We deployed a special wedding landing page on Boesen's WordPress based site that contained an Aweber sign-up form and a welcoming video. The auto-response to the form delivers a download link to receive the free Boesen wedding planning ebook containing 52 weeks of recommendations, suggestions, and planning information.

Then, over the next 52 weeks, e-mails are dripped weekly with enhancements to the e-book. For example: let's say the chapter is on honeymoon planning. The follow up e-mail that arrives will offer additional videos, audio, and enhanced content on destinations, air travel, and hotel review sites.

The form also has a phone number box for those who wish an immediate connection with a floral planner.

Results: Boesen's inbound leads specifically for weddings increased by 5 per week. Not much you say? Consider that the average wedding package (across the board including DIY $200 packages and $10,000 mega-packages) is about $1400. Of the 5 new leads per week, 3 were offering up their phone number. I must add that if a bride speaks with Tom Boesen personally to plan her wedding flowers, she will decide to work with him 9 out of 10 times. In sales speak, Tom's close ratio is over 95%. Remember, "There's no TOM behind their dot com!"

So the breakdown is:

- 5 new leads per week
- 3 provide phone number and turn into live appointments
- 2.7 (90%) of those become sales
- $1400 (avg. sale) x 2.7 (# of sales/week) x 4 (weeks) = $15,120/mo in additional revenue from the e-mail drip campaign

I can tell you that this campaign didn't cost anywhere near that much to produce. As a MultiThreader, I used 4 people to execute simultaneously on this project (Multi-Threading!) while I managed the project via collaboration software with total transparency and visibility for the team and the client.

- Resource 1: e-book creation and formatting including stock imagery
- Resource 2:   WordPress landing page technology tweaks and layout
- Resource 3: Graphics and buttons for the landing page
- Resource 4: E-mail campaign and e-mail content set up

Compelling.   The full Boesen the Florist case study is included in section 4.

MultiThread Marketing Lesson from AOL

How many discs did you get in the mail? 20? 50? I know some people created ornaments, wind chimes and even wall decorations with extra discs. Yet, it was an opt-in process (you chose to sign up or sign in). And it wasn't a coincidence that once you signed up, the discs slowed up — until you stopped using the service. Then, they ramped up again.

Now this was before everyone had a computer. Before everyone used e-mail. Now everyone has some sort of device and everyone has e-mail Unbelievably, some still use an AOL e-mail address - but not you, right?

Here's my point: Keep things opt-in and "ping" them in a valuable way a couple of times in the first few weeks. Get them in the habit of receiving valuable information from you via e-mail. Don't make them feel like it's spam (unless it is) - they opted in.

CHAPTER 21

Video Killed the Marketing Star

"Fly to Des Moines and we'll shoot your commercial in my basement."

That was my suggestion a couple years ago and it worked great. Where else could you put up a green screen studio for about $15 in Walmart material sewn together, a $225 light kit, and a consumer grade camera?

YouTube.com is considered the "Number 2 search engine behind Google" if their traffic is broken out from the mother ship. (1) After you pick yourself up from that one realize that the MultiThreader must understand video enough to produce "good enough video" without breaking the marketing budget. Let's break down some key areas in the video equation.

The Search Heard Round the World

Take a peek at this example of the power of video:

1. I searched YouTube to learn more about a topic that our clients find vexing and painful yet important.
2. I found 2 great and informative videos done by a company that really opened our eyes. The videos were effectively narrated PowerPoints.
3. We watched the videos during an all-hands meeting.
4. During that meeting I got the idea to create a short video that really highlighted the pain this topic causes and VERY briefly how our solution addresses it. This wasn't a sales pitch but it had about 10 seconds of powerful messaging that we ARE the solution.
5. I created a blog post with that video embedded for people to refer back to and share which is automatically tweeted out through our account http:// twitter.com/birddog_jobs
6. People and companies have searches and alerts set up so any time the key terms this video addressed is mentioned, they're alerted, and one firm specializing in the issue we addressed found us, followed us on Twitter and re-tweeted our blog post containing our wacky video.
7. Turns out the firm that retweeted our post follows about 130 related firms and consultants that we now follow (anytime you follow someone they are alerted and the "follow backs" have been flowing in.)
8. I sent an e-mail to the company that retweeeted our video post and explained who we are, that we loved their videos and found them tremendously valuable, and asked if they wouldn't mind opening a dialog to see if we can add value to each others' networks of clients.

9. We had the meeting and discussed our target markets, our solution, and that we typically deal w/companies that are 250 employees or less and clients that haven't used a solution like ours in the past.

10. Their company representative said "We refer people to systems like yours all the time and the most under-served segment of the market is the sub-100 employee space."

11. We had a demo a few days later that resulted in that company's entire management team agreeing that, between 50 and 100 of their current clients needed this solution, and that this would round out our offering so we needed to look at a private or white label type offering. We've had multiple meetings and co-presentations to date.

12. This has become a 6-figure sales pipeline opportunity and we've received one referral already from the company.

13. We've signed an agreement with the firm and are integrating them into our software officially as a partner.

14. This partner's knowledge has given my firm the ability to rework some existing data we capture into a highly marketable "product."

Let's summarize:

- Searched on YouTube and found relevant material to share and promoted it socially and on our blog site.
- Rapidly created fun video using flip cam and iMovie software that was pushed out via blog and social tools.

- Established instant rapport and credibility with complementary company and approached with an investigative "see if we can add value to each other" attitude.
- Looking at large caliber deals with big upside.
- Total investment on my part as MultiThread Marketer?

8 Hours.

CHAPTER 22

Boesen the Florist

Following is the case study we presented to Boesen the Florist in February of 2010. I expect an update to this case study in late 2011.

I recommend that before reading this chapter, you visit http://bit.ly/b3Cjcl and grab the actual case study for the images, charts, etc. No opt-in is required and this link points directly to the PDF.

Setting the Stage

Boesen the Florist is a locally-owned florist in Des Moines, Iowa. The owner Tom Boesen was looking for additional incremental sales and wanted to gain brand credibility with younger clientele who didn't necessarily grow up with the Boesen brand as a household name.

When we first began collaborating with Boesen on using new media and iInteractive marketing tools, admittedly their team was skeptical.

The big question loomed over our potential engagement:

"With so many new tools and strategies to explore., Twitter, Facebook, Youtube, etc., how can one actually use these tools in a business context without wasting time, AND who's going to do it?"

Our team explained that these new media marketing methods are part of a comprehensive web strategy and that this is where 99.99% of companies go wrong. *Most companies bolt on additional new media marketing tools without having any idea of what their outcomes, goals, or metrics should be.*

Boesen was well aware that the face of marketing was evolving around its 86-year-old brand, but it hadn't yet found the right catalyst for change. Boesen is a pioneer in the floral industry and a technology leader. Think back and you'll recall that the floral industry was one of the very first to embrace e-commerce fully and go online with stores. But the pace of change in the market and the explosion of social marketing had outpaced Boesen's ability to adapt.

The Boesen team did not know then that redirecting some marketing dollars to give this social media stuff a try would have such a positive change on its brand, public awareness of its philanthropy, and its sales revenue.

What We Discovered

Our team dove in and discovered that like so many in the floral industry, Boesen's web presence was very powerful as a shopping cart and back end ordering and inventory system. It was not however, conducive to putting out compelling and fresh marketing content. The structure of the site and most sites for that matter are more static in nature, and doesn't have the capacity for new material (video, audio, text) to be added on the fly by anyone with the ability to type in a word processing program. Additionally, Boesen did not have the staff to create the content or the expertise to execute on things like basic video editing.

Boesen needed a foundation for its own marketing content distribution. That foundation is an additional marketing site or micro site built on the content management solution called WordPress.

The bottom line is that WordPress provides most companies their best shot at being found online through search engines and it provides an easy interface for rapid content release.

As part of its overall web strategy, Boesen also needed to:

- Learn how to listen online to take advantage of positive chatter in the online marketplace while addressing any negatives head on and proactively.
- Learn how Twitter can be used as a community-building tool where people spread brand awareness, take advantage of offers, and become brand evangelists for you and your company.
- Learn how to leverage the power of engagement through a Facebook Fan Page.
- Learn how to engage the 18-40 year old demographic by actually participating and speaking WITH them.

- Learn how to use video to give eye popping floral arrangements the "3rd dimension".
- Learn how to transform and leverage an 86 year old brand, famously known among the over 50 crowd into a brand that younger "digital natives" can embrace.
- Learn how to use the special online offers that limit shopping to a couple highlighted items out of an overwhelming number of choices.
- Learn how to use analytics to understand the value (or lack thereof) of internet banner advertising.
- Learn how to leverage multimedia to show off the PEOPLE that make up the company since Boesen is ultimately a collection of people representing the Boesen brand.
- Learn how to get Boesen staff involved and committed in the process of developing a robust online brand.
- Accept that a firm like createWOWmedia had the expertise and ability to assimilate into Boesen culture so we could kick start its online efforts by being a proxy for content creation.

This is a big area of controversy for many. We don't advocate merely ghost writing for CEO's. However, we do believe that it is possible and desirable for the right company or individual to create content for another firm that supports its brand and mission. It is critical however that your company chooses the right partner with the ability to assimilate into the culture and learn a company's processes and business model quickly.

Our team had a lot of work to do. We were also facing a lack of good tracking data and analysis on what has been happening with the site, orders, behaviors, etc. This is not uncommon. It's very difficult to keep one's eye on the analytics ball when leadership is focused on a shrinking economy, promotions, human resources, etc.

More simply stated, createWOWmedia's goal for Boesen was to create a comprehensive web strategy using interactive media that would integrate into and modify Boesen's overall marketing strategy resulting in increased online sales and a newly engaged buying demographic of 18-40 in the Des Moines metropolitan area.

Twitter

createWOWmedia began driving Boesen's Twitter Account around mid-October 2009. During that half-month we were focused on building a strong hyper-local following. During the first full month, we built the following base from 60 to 392 followers. Since that first drive to build the base we've seen fantastic growth.

Here is Boesen's initial three month historical following trend on Twitter (Nov '09 - Jan '10): While quantity of growth has slowed, as it always does, it is still on a strong upward trend. You will see that since October Boesen's Twitter following has grown to 573 and based on these trends they could reach 620 in the next 30 days, not accounting for any Valentine's Day boost. They actually hit 641 and are beyond even that.

We utilize Twitter to accomplish a variety of things.
First, we have maintained conversational contact with our following so they know Boesen isn't just trying to "spam" them with sales.

Second, we've successfully had contests and give-aways that reinforced the Boesen following and even caught attention in the Des Moines Register. Boesen the Florist was included (through linking and embedding of videos) in other blogs and videos which resulted from this type of online participation.

We've also learned a lot about the limits of the participation of the Boesen "fan" base. They are mostly self-interested when it comes to contests and events and if they aren't personally getting something out of it or it requires too much work for the prize they won't do it. This enabled us to adapt and offer new contests that are still attractive but original.

Third, we've sold through Twitter. From October going into January, Boesen was approaching 4,000 click throughs. Approximately one half of those went directly to Boesen.com. Most of the other half went to FlowersDesMoines.com.

The spikes and drops are based on content. Days with great deals and a lot of conversation have more click throughs. Participation is key.

To date, we have not even fully leveraged Twitter for Boesen's benefit, meaning, there's still more that can be done for even greater success. For example, when they received loads of mini-carnations by mistake and the grower didn't want them back and just gave them to Boesen, we sold them in bunches for $1.00 to move them quickly and make some extra cash. That had a huge response online and people flocked in. Not only did it result in some extra revenue but it again reinforced Boesen as the florist that cares about their customers with such great deals.

Twitter is a vehicle that can facilitate a change to strategy within minutes. If inventory is too high because of an overbuy on something or a day is abnormally slow we can do quick special offers to drive business and boost those slow days.

Facebook

Boesen's Facebook Fan Page was constructed in September and had a trickling of 10 fans when Boesen engaged createWOWmedia to drive its online efforts in mid-October. After a little work we quickly got the Boesen fan base to break 200 and within four months they had 273 fans of Boesen the Florist on Facebook.

During this period they also lost 20 fans which is natural. Fan pages are a challenge to grow. They require a certain balance of new content while not becoming overbearing. A company can quickly lose their base on Facebook by posting too frequently.

Boesen has had a steadily growing fan base in the very demographic we were targeting. Seventy two percent of their fans are 34 years old or younger. Reaching this group can be a daunting task but we managed to do it with ease while still providing an environment for those over the age of 34. Sixty-seven percent of their fan base is female which can be expected.

We've utilized Facebook in a couple of ways to support Twitter and FlowersDesMoines.com. We've integrated these tools to create a powerhouse of online engagement. All of our online contests and events were simultaneously run through Facebook and our micro site.

We've also created an open environment where individuals can and have uploaded their own photos of flowers from Boesen the Florist. Facebook will continue to be a strong part of the Boesen online brand. Some people choose Facebook exclusively as their home base of online social marketing activity so it's not wise to ignore this market. Facebook has the most robust and accessible marketing and demographic data as well. We took a small dive into Facebook ads during Valentine's Day and received incredible results. For a minimal cost we were able to get almost 700,000 impressions within 25 miles around Des Moines, Iowa in three days during the peak buying period before Valentine's Day.

FlowersDesMoines.com

FlowersDesMoines.com is a micro site that supplements Boesen.com. FlowersDesMoines.com is used as the content driver of your online campaign. Content production on this site also started around mid-October. Since then we have published 123 different articles and have been quoted and referenced in numerous news articles and blogs. That is huge. In a mere five months (Oct '09 - Mar '10) FlowersDesMoines.com was already providing up to 15% of Boesen.com's total monthly traffic which is astounding.

In just a matter of months our monthly traffic was equal to 33% of what Boesen.com received on a monthly basis.

Most traffic is derived organically through search engines. Our constant output of new content with strategic keywords has really boosted Boesen's online findability or Search Engine Optimization (SEO).

The site has been keyword focused since inception. Even the URL, FlowersDesMoines.com, was chosen to make Beosen's site more findable. Within a week this site (with almost no content) was on the first page of Google results. Our focus on the keywords, Boesen, flowers, des moines, iowa and boesen the florist has paid off. Here is a chart showing your top ten keyword searches.

The majority of these visitors are there for the first time. Actually, FlowersDesMoines' average new visits per keyword is 94.66%. The incredible thing we've found is that only one of these keywords actually has the business name Boesen in it. In contrast with Boesen.com where virtually every major keyword has the word Boesen. This further proves we've been able to hone our efforts into very specific targets without wasting reach or cannibalizing traffic that would have already gone to Boesen the Florist. We've brought in traffic that likely wasn't predisposed to purchase from Boesen the Florist because they didn't search for Boesen. To take advantage of this traffic we've fully integrated with Boesen to increase referrals to their catalog site. We've also taken it further and actually started selling certain items through FlowersDesMoines.com as well. To date, FlowersDesMoines.com has a higher conversion rate than any other referal sites and even higher than any of the search engines including Google.

This micro site has been used to promote events, contests, give-aways, specials and much more. We've successfully integrated it with Twitter and Facebook so that we are able to maintain and grow the online energy that has been cultivated.

Online Community

It's apparent Boesen now has a loyal community online. These individuals have expressed their love of Boesen the Florist. They participate in well-planned contests and events. They take advantage of deals. When someone complains or asks generally about florists in the Des Moines area online, they are the first to stand up and shout that Boesen is the best. They are not shy about it. However, they also don't like paying for delivery. Most of our followers online come into the store to take advantage of deals.

That's why all our online specials have allowed for cash and carry pickups. Boesen is a local shop so it's not troublesome to get over to one of the locations and pick something up.

$$$ Results

Due to the nature of the Boesen following it's impossible to show real statistical growth online because most of them aren't making purchases online. Even so, we have seen growth online. Unfortunately we don't have more than three months of history working with Boesen's online campaign. However, going back through our three month history online sales compared to the previous years have been up during a time when most floral shops and retail in general are flat or down. Here's the data:

Month   Growth
January   3.41%
December   1.04%
November   14.19%

There is growth online and even in-store. When looking at these month-to-month growth figures it is vital we remember that for every online order there were likely three additional orders that were derived from your online marketing efforts. This translates into much more than even the 1.04% growth from November to December. Not only has there been online growth but Boesen also finished 2009 with positive growth over 2008 when the economy was in far better shape on average for the year.

Metrics

The hardest part is showing a true ROI for interactive media marketing. So here's how we've done it:

To begin, website traffic is easy to track. We utilize a system that allows us to measure all the traffic, traffic sources, conversion, keyword's success rates, and much more. However, we realize that most people who buy flowers want to talk to a florist. Furthermore, as we previously stated, most of Boesen's online following is local and does not want to pay for delivery. This makes tracking even more difficult.

To get a more realistic perspective, we've used the industry accepted benchmark that for every one Internet order there were three phone orders that were derived from online marketing efforts. This metric gives us perspective. From here we can start to build a more accurate depiction of how effective these marketing methods are and the ROI they provide Boesen. We've started using Google Voice. Google Voice is a free VoiP phone number you can receive. Right now they are still in beta and are by invite only. However, if you can get your hands on multiple local numbers you're set.

What we do then is this. Boesen has numerous toll-free numbers they are allocated for use at no extra charge. We then split up these toll free numbers among the different Internet marketing tools we use. Each one of these numbers than forwards to a Google Voice number which is forwarded to Boesen's main number. This does two things.

1. Google Voice doesn't have toll free numbers, so to avoid our customers ever having to pay to call us we've placed Boesen's current allocated toll free numbers as the first forwarding numbers.
2. We then forward them to Google Voice because Google has better analytics than a toll free phone number bill.

Furthermore, we wanted to keep a local number as well because Boesen's customers like that. So they can call at either point A on the toll free number or at point B on the Google Voice number. We then can work with another industry-accepted benchmark that every phone call longer than 4 minutes is likely an order.

Now we can gain a much more accurate picture of the impact and ROI of our interactive media marketing efforts. Not only can we compare Boesen with the industry standard but we can work to raise the bar and make Boesen the new benchmark for the industry.

For Facebook and Twitter, we use other application tools that allow us to track click-through rates as well and based on our average online conversion rate, we can derive ROI from those sites as well.

When it comes down to it, social media marketing can be measured no matter what you've heard elsewhere. You just need to know how to do it.

Conclusions and Future Plans

Even though we have seen great success developing Boesen's successful and profitable Interactive Marketing campaign, the work is really just beginning.

 Like most companies, Boesen the Florist doesn't have a full-time Interactive Marketing Director and budgets must be adhered to so we're deploying new tools and methodologies incrementally.

Our content production and online monitoring strategies are maturing, so we're able to explore areas that have barely been touched like Pay-Per-Click ad campaigns and in-store conversion tracking through the POS system or via a Boesen Associate.

We've also barely begun sharing the true beauty of the products created by Boesen's amazing design staff. Most images on the Boesen.com site are more stock in nature.

We will soon have a streaming web cam in the design studio so brides and others can see their arrangements being assembled in "real-time".

We have niche markets that remain untouched too. Currently, we're building out a hyper-targeted content portal that will serve a massively underserved market in the Des Moines metro. This portal will leverage multimedia and high value content to drive sales and brand equity.

Boesen the Florist realized early on in our engagement that they could not execute on what we laid out in our strategy plan. They had to have ongoing help. Without this piece of the puzzle, this entire plan would likely have been considered a failure. It is wise for companies considering an Interactive Marketing campaign to take note. It's not good enough to simply "put Jenny on this new media stuff because she just graduated with a marketing degree". Effective online strategy takes multi-phase and multi-variant execution that most firms will find overwhelming, i.e., setting up systems that will support content, then creating the content, then editing the content into something powerful, etc.

As we progress with Boesen the Florist, I'm sure we'll provide addenda to this living breathing case study. We continue to learn in our engagement every day.

It's an honor and privilege to be charged with renovating a brand that's been around for over 80 years old brand for a company that's an institution in the state of Iowa. We definitely handle that with care.

Special thanks to Tom and Frank Boesen and Steve Schultz for their massively important input and creativity throughout the last five months. We can't do what we do without your confidence, support, and funding.

CHAPTER 23

Section Four: Additional MultiThreading Resources

In this section I have provided some tools you can use in your quest for MultiThread Marketing Rockstars. If you are rockstar, then study up on the questions and scenarios in the following pages so you can make better choices about where you choose to work.

CHAPTER 24

Days In the Life of the Modern Marketer

What Would You Say Ya...Do Here? (The Bobs to Symkowski in "Office Space")

Over time, I will be publishing "day in the life" videos and podcasts to give you a real flavor for what it's like in the MultiThread marketing trenches where I function every day. I think this delivery method will be more intriguing and valuable, especially for those wanting to move into marketing leadership.

You will find these videos, audio, and other podcasts and written pieces on my personal site at www.douglasE-mitchell.com.

CHAPTER 25

The Misguided Marketing Job Description

I have an exercise for you. Go on any job board and search for the keyword "Marketing" and note what you find. In many cases I find sales jobs cloaked in the "marketing" moniker.

The description that gives me the most heartburn is "Director of Marketing and Sales". The depth and breadth of responsibility in many of these descriptions is monumental and often a set up for failure. Answer this question: If sales are down, how much time do you think the executive team and board of directors want you to spend on marketing? Leadership will default to "Just close business!" and your true marketing efforts will suffer.

Here are few key things to remember when you are creating your marketing job description:

- **Packing in 20 line items covering a massive range of responsibilities and business areas should make candidates very nervous about your expectations.** Many descriptions I read look like 5 jobs wrapped up into one position. Face reality and don't set the bar at unachievable levels.

- **Think through what you expect the candidate to do, and what you're comfortable outsourcing.** If you simply will not stand for using an outsourced copy writer for example, say so.

- **Don't just think about the salary and benefits package for the candidate.** Consider the outsourcing budget that he or she will need to fulfill the role. As the VP of Marketing he might develop a strategy, develop the tactical plan, and then leverage external resources amounting to 40% of his salary. Will you feel cheated because he's not doing more on his own? Confront your own personal biases and predispositions so you don't resent the new hire's ability to free up his time to think, plan, and learn as opposed to purely executing.

- **Be realistic about the autonomy and self-guidance you're willing to give your new marketing leader.** Nothing is more frustrating for you or your candidate than saying "Create the strategic marketing plan and run with it!" only to be cut off at the knees after a few weeks because the management team wants to go with their gut. If you're not willing to roll with your Multi-Threader's plans, tell them upfront.

- **Clearly identify all of the areas in which you'd like "good enough" knowledge.** Outline all the cool stuff you need the candidate to do like:

1. Write website or blog content (http://blog.birddog-jobs.com)
2. Create video using green screen or in house and edit with imovie or finalcut - or whatever you bring on a laptop.(www.youtube.com/user/birddogjobs)
3. Write scripts and record audio segments in our own in-house studio.
4. Use WordPress to adapt rapidly online (create new pages, new content, etc.)
5. Do graphical work using Adobe CS or open source alternatives.
6. Do e-mail marketing using Aweber and/or Mail-Chimp.
7. Do "light" html work including interfacing with coder types to make web pages on our traditional html sites.
8. Communicate frequently with other PR and marketing professionals.
9. Do market research by calling existing, new, and lost customers to help us determine how to be better, stronger, and faster in the market.
10. Leverage social tools like Twitter, Facebook, and LinkedIn to build business and create value for others.
11. Have a ton of fun running full speed and accomplishing 12 times more than most companies EVERY SINGLE DAY.

When building your MultiThread Marketing job description, just remember to be honest. Nothing is more painful or expensive for both parties than lack of clarity at this stage in the hiring process.

CHAPTER 26

The Marketing Interview

I am not an HR professional nor am I an attorney so please, do whatever you need to do and ensure your questions are legal, ethical, and proper. Here are some examples of what I'd like to know if I'm trying to uncover the MultiThreading talents that win.

These are questions to uncover the candidate's uses of resources, their perception of what constitutes a value, and their connection to resources.

1. Do you know freelance graphic designers in town? At what rates? How many revisions does it typically take to achieve your desired outcome at each rate? Do you have examples of their work? (This is designed to uncover the marketer's network and reach.)

2. Have you been in control of discretionary marketing budgets before? What size? How did you utilize your resources? (Spending money strategically isn't as easy as it sounds. If you plop someone who's had a $60k

budget into a role with 1.2 million available at her discretion, this can be a challenge you'll want to address.)

3. Have you leveraged advertising or marketing agencies in the past? What was your experience like? (Agencies aren't all bad but if the marketer is used to spending $10k per month with a $150 per hour agency and spending $5k for a testimonial video shot in a glamorous studio, odds are he will not appreciate your gritty trench warfare.)

These questions uncover the candidate's distinctions between leadership and execution.

1. Describe a "great day" in your marketing life. (This question can be a key indicator of whether your marketer will always be buried under a mound of tasks, or leading and executing a much larger reaching strategy.)

2. How comfortable are you delegating tasks like design and content creation? (If you hire someone who's passion is design, he will likely spend a lot of time perfecting the swoosh so to speak.)

These questions uncover a candidate's core beliefs about marketing and what their mission and purpose will be at your firm.

1. What's your definition of marketing? (No perfect answer here but you may get a real sense of whether a candidate is focused on "looks" instead of "driving demand".)

2. What's the purpose of a website? (Does the candidate view your site as a brochure or a demand generation conversion machine?)

3. Let's say we do get a visitor to our website. What should happen next? (This question will uncover the candidate's understanding of conversion marketing, i.e., getting the visitor what she wants so she will become a lead, buy our product, etc. versus merely "looking around.")

These questions will uncover the candidate's technological and multimedia skill set.

1. What blogging platforms have you used?

2. What social platforms do you actively use to market? (Throw in the "top 5" social tools or ones that your firm already uses to seed the conversation.)

3. Which online collaboration tools have you used to manage internal or external resources?

4. Have you done any podcasting or video work? What equipment set ups have you used and what do you prefer? (Between GarageBand and a $100 Blue Yeti microphone your marketer will be a powerhouse.)

5. Have you ever done green screen video on your own? (You're not looking for a video professional here. You want familiarity with shooting green and knowing that it's easy.)

6. What are some content outposts you typically use to push your message out?

7. Do you consider yourself a dabbler, proficient, or expert using [fill in the blank software or hardware platform]?

CHAPTER 27

That's A Wrap

I hope you enjoyed this book and will use it as a catalyst to expand your marketing horizons through Mutli-Threading. Technology is expanding at such a rapid pace these days, and if a marketer doesn't have an amazing cast of "threads" around, it's easy to fall behind. Also remember that leadership, passion, and execution trump technology on most days so don't get caught up.

I stand on the shoulders of many amazing men, women, and software applications every day to enable my success. You can easily do the same.

Welcome...to MultiThreading.

All the Best,

Doug

www.douglasEmitchell.com

CHAPTER 28

References

Here are some sites and resources I used in building this book:

Preface

1. http://www.webopedia.com/TERM/M/multi-threading.html
2. http://en.wikipedia.org/wiki/Instructions_per_second

The Basics: Seriously?

1. http://en.wikipedia.org/wiki/Mavis_Beacon_Teaches_Typing

Watcha Been Readin' Lately?

1. All of the books recommended in this book are available at my Amazon Store - **http://bit.ly/marketing-booklist** *This is an affiliate link and I will make a tiny bit of money if you purchase this route vs. ran-

domly visiting the links yourself.   If you'd like to deny me this simple pleasure, please visit Amazon.com.

Here are some titles I highly recommend:

- Inbound Marketing - Brian Halligan and Dharmesh Shah
- Six Pixels of Separation - Mitch Joel
- The New Rules of Marketing and PR - David Meerman Scott
- Socialnomics - Erik Qualman
- The Shallows - Nicholas Carr

Measure Twice - Cut Once - 1300 X Per Month

1. http://voices.washingtonpost.com/44/2008/11/20/obama_raised_half_a_billion_on.html

Paying Less for Organic

1. http://en.wikipedia.org/wiki/AdWords

Video Killed the Marketing Star

1. http://www.comscore.com/Press_Events/Press_Releases/2010/1/com-Score_Releases_December_2009_U.S._Search_Engine_Rankings

I'm Sorry Dave But I'm Afraid I Gantt Do That

1. http://bit.ly/famhby - Central Desktop Trial Offer Affiliate Link.   If you don't want me to get the few bucks, visit CentralDesktop.com on your own.

How's the Candidates Google Juice?

1.   http://bit.ly/aC43j5

Published by FastPencil
http://www.fastpencil.com